Methods and Techniques in Nematology

Authored by

Ebrahim Shokoohi

*Department of Biochemistry, Microbiology
and Biotechnology, University of Limpopo
Private Bag X1106, Sovenga, 0727
South Africa*

Methods and Techniques in Nematology

Author: Ebrahim Shokoohi

ISBN (Online): 978-981-5313-68-0

ISBN (Print): 978-981-5313-69-7

ISBN (Paperback): 978-981-5313-70-3

First published in 2025.

need for a court order if at any point you breach any terms of this License Agreement. In no event will any delay or failure by Bentham Science Publishers in enforcing your compliance with this License Agreement constitute a waiver of any of its rights.

3. You acknowledge that you have read this License Agreement, and agree to be bound by its terms and conditions. To the extent that any other terms and conditions presented on any website of Bentham Science Publishers conflict with, or are inconsistent with, the terms and conditions set out in this License Agreement, you acknowledge that the terms and conditions set out in this License Agreement shall prevail.

Bentham Science Publishers Pte. Ltd.
80 Robinson Road #02-00
Singapore 068898
Singapore
Email: subscriptions@benthamscience.net

BENTHAM SCIENCE

CONTENTS

FOREWORD I i
FOREWORD II ii
PREFACE iii
ACKNOWLEDGEMENTS iv

CHAPTER 1 INTRODUCTION 1

CHAPTER 2 SAMPLING 4
    GENERAL CONSIDERATION 4
    TOOLS FOR SAMPLING 5
    SAMPLING FROM AQUATIC AND WETLAND 5
    SIZE OF THE SAMPLING AREA 5
    SAMPLING DEPTH 9
    SAMPLING PATTERN 9
    CONCLUSION 12

CHAPTER 3 NEMATODE EXTRACTION 13
    EXTRACTION OF SOIL AND ROOT NEMATODES 13
        Tray Method 13
        Sugar Flotation Method 15
        Incubation Method 16
        Sieving Method 17
        Dissecting Method 17
        Seinhorst Cyst Extraction Elutriator Method 18
        Baunacke Method 19
        Maceration and Filtration or Blender Method 20
        Wood and Compost Extraction 20
    EXTRACTION OF ENTOMOPATHOGENIC NEMATODE (EPN) 21
        Nematode Isolation 21
        Practical Procedure 21
        Investigation on Parasitic Nematode of Mosquito 23
        Practical Procedure 23
        Mosquitoes Preparation in the Laboratory 23
        Propagation of Mermithid Nematodes 24
        Evaluation of Different Extraction Methods 25
    CONCLUSION 28

CHAPTER 4 NEMATODE OBSERVATIONS 29
    FORMALIN-GLYCERINE METHOD 29
        TAF (Triethanolamine Formalin) Method 31
        Slide Preparation 32
        Benefits of Temporary Slides 32
        Benefits of Permanent Slides 33
    CONCLUSION 35

CHAPTER 5 NEMATODE MORPHOLOGICAL OBSERVATIONS 36
    MICROSCOPE JUSTIFICATION 36
        Microscope Types 36
        Calibrating the Eyepiece Graticule 38
        Calibrating the Microscope for Drawing 39
        Maintenance of microscopes 41

Differential Interference Contrast (DIC) 41

NEMATODE MEASUREMENTS 43

Pictorial Measurement Guide 45

SCANNING ELECTRON MICROSCOPY (SEM) 45

Preparation of Specimens for SEM Study 55

CONCLUSION 57

CHAPTER 6 MOLECULAR DIAGNOSIS 58

DNA EXTRACTION 58

Isolation of Nematodes 58

DESS Solution 59

How to Transfer Nematodes to Ethanol 59

Genomic DNA Extraction 59

DNA Extraction Using the Chelex Method 60

Isolation of DNA from Individual Nematodes and their Connected Bacterial DNA 61

DNAzol kit DNA Isolation Protocol for Individual or Pooled Nematodes 62

NaOH Digestion of a Single Nematode 63

Nucleic Acid Extraction using Lysis Buffer I 64

DNA Extraction from Glycerine-Embedded Nematode Specimens 64

Lysis Buffer Method 65

COMMON MOLECULAR MARKERS 65

Sequence Characterized Amplified Region (SCAR) 65

Primers (Case Study: *Meloidogyne*) 68

Restriction Fragment Length Polymorphism (RFLP) 68

Preparation of DNA Digestion Master Mix 70

Post-digestion Steps 71

Amplified Length Fragment Polymorphism (AFLP) 71

Polymerase Chain Reaction 73

PCR Components 75

DNA Template 76

Quality of DNA Template 77

Evaluating the PCR Products 78

CONCLUSION 80

CHAPTER 7 BASIC BIOINFORMATICS IN NEMATOLOGY 81

GENERAL INFORMATION 81

CHECKING THE SEQUENCES 82

BLAST OF THE SEQUENCES 86

ALIGNMENT AND PHYLOGENETIC ANALYSIS 89

WHAT SEQUENCES SHOULD BE SELECTED? 89

ALIGNMENT USING MEGA 90

COMPUTING PAIRWISE DISTANCE 91

CONSTRUCTING TREES 92

Maximum Likelihood Tree 92

Advantages 92

Disadvantages 94

FEATURES OF A PHYLOGENETIC TREE 94

BOOTSTRAPPING 96

OUTGROUP OF PHYLOGENETIC ANALYSIS 96

ONLINE TOOLS FOR NEMATODE IDENTIFICATION 96

CONCLUSION 97

CHAPTER 8 BIODIVERSITY ANALYSIS 98
  CONVENTIONAL BIODIVERSITY 98
    Sampling 98
    Evaluation of Soil Samples 99
    Counting of Nematodes 99
    Biodiversity indices and data analyses 100
      Shannon index 101
    Simpson's Index 101
    Richness 101
      Evenness 102
  MOLECULAR BIODIVERSITY 103
  CONCLUSION 108
CHAPTER 9 NEMATODE REARING AND GREENHOUSE STUDIES 109
  CULTURING OF NEMATODES 109
    General consideration 109
    Globodera spp. 110
    Heterodera spp. 110
    Meloidogyne spp. 110
    Radophulus spp. 111
    Pratylenchus spp. 111
    Aphelenchoides and *Bursaphelenchus* spp. 111
    Mylonchulus spp. 112
    Panagrolaimus spp. 112
    Caenorhabditis spp. 112
  GREENHOUSE EXPERIMENTS 113
    Nematode Inoculum 114
    Greenhouse Technique for the Evaluation of *Meloidogyne* 114
    Rearing *Meloidogyne* Males 116
    Reproduction Assessment 116
  CONCLUSION 117
REFERENCES 118
SUBJECT INDEX 128

FOREWORD I

Nematodes, also known as roundworms, are essential to soil ecology and play a crucial role in maintaining soil health. These organisms have various feeding groups, including bacterivores, fungivores, herbivores, predators, and omnivores.

Despite the herbivores or plant-parasitic nematodes, many of them, such as free-living nematodes, are beneficial for agriculture and crop production as they help in breaking down organic matter, such as dead plant material and animal waste, into nutrients that plants can absorb. In conclusion, nematodes are more than just tiny worms in the soil. They are critical to the health and productivity of agricultural ecosystems and play a significant role in ensuring global food security.

"Methods and Techniques in Nematology" is an excellent book for anyone interested in the nematology discipline. This book is specifically designed to help lecturers, researchers, farmers, and students deal with nematode problems. The book contains a variety of techniques with detailed explanations and high-quality photographs to make the learning process easier and more engaging. These photographs help bring the topic to life and make learning more enjoyable. You will be able to see the nematodes up close and appreciate their unique features. You will find all the essential information you need to understand nematodes and learn how to address any related problems. The book covers various techniques, from conventional to molecular, particularly for those wanting to start with nematology, all of which are explained in simple and easy-to-understand language.

In conclusion, "Methods and Techniques in Nematology" is an excellent resource for anyone interested in nematology. It is an easy-to-use practical guide that will help you understand nematodes and learn how to address any related problems.

Prof. Ebrahim Shokoohi's book on nematology is an invaluable academic resource that provides students with a comprehensive understanding of the subject. With years of experience teaching at various educational levels, ranging from BSc to Ph.D., Prof. Shokoohi's book is a reliable source that would appeal to a wide audience seeking a deeper understanding of the subject matter; I feel he accomplished it brilliantly.

Zafar A. Handoo
USDA, ARS, Mycology and Nematology Genetic Diversity
and Biology Laboratory, BARC-West
Beltsville, MD 20705, U.S.A.

FOREWORD II

Nematodes are remarkable organisms with amazing abilities that have made them the most common animals on the planet; however, they are rarely seen because they are microscopic. Yet people who observe them often share a common bond and instant friendship. My relationship with Dr. Ebrahim Shokoohi is just that, a brotherhood of camaraderie that is based on our mutual admiration of all things nematode. We have worked together describing the wonderful world of nematodes.

Because nematodes are so small, they are difficult to work with. For this reason, they are often neglected, but with proper techniques, they become extremely important in many different types of studies. Fortunately, Dr. Ebrahim Shokoohi has made an effort to assemble various methods of handling and investigating nematodes. They range from sampling and separating them from soil, making slides and examining them with a microscope, designing greenhouse experiments and analyzing the results, to extracting their DNA for taxonomic studies.

This book contains valuable information for anyone interested in working with nematodes because it clearly illustrates many techniques that are described in a logical, step-by-step manner, which makes it easy to follow. The nematology community will be very happy to have this valuable resource for their use.

Jonathan D. Eisenback,
Professor of Nematology
Former President of the Society of Nematologists, Virginia Tech
Blacksburg, VA 24061, U.S.A.

PREFACE

The research on nematology is significant for researchers, students, and everyone interested in this science discipline. While I was teaching nematology for about ten years, I came up with the idea of documenting the methods and techniques that would be useful for everyone. In this book, I have presented the relevant methodology within a conceptual framework of different scopes within nematology that renders technical information that is needed for students and researchers. Methods ranging from sampling to advanced techniques, including molecular surveys, are discussed in this book. The methods are presented in a way that is adaptable for the students to use in formal courses, which can also be functional when used daily by academics and educational institutions. In almost part of the book, the author's experience and the available knowledge of the expert in nematology create an opportunity to easily run the experiments and surveys.

Ebrahim Shokoohi
Department of Biochemistry, Microbiology
and Biotechnology, University of Limpopo
Private Bag X1106, Sovenga, 0727
South Africa

ACKNOWLEDGEMENTS

I am grateful to "Allah" for blessing me with a healthy mind to think about the creations and learn from nature. I am thankful to all my teachers and professors for all I have learned in nematology and plant pathology. I would like to thank Prof. Annette van Aardt for revising this book. I would also like to thank Mr. Panahi for providing some of the high-quality pictures for this book. I would especially like to thank my lovely wife for the excellent atmosphere and courage to finish and publish this book. Finally, I would like to thank my parents for the endless support, enthusiasm, and love they have given me. I dedicate my book to my beloved, my wife and son, Adrian.

Ebrahim Shokoohi
Department of Biochemistry, Microbiology
and Biotechnology, University of Limpopo
Private Bag X1106, Sovenga, 0727
South Africa

CHAPTER 1

Introduction

Nematology is an important branch of biological science that focuses on the study of a diverse group of roundworms known as nematodes. In addition, nematology plays a crucial role in agriculture. It is integral to the management of crop pests and the enhancement of agricultural productivity through the study of nematode interactions with crops. Additionally, in the medical field, nematology is vital for controlling diseases caused by parasitic nematodes. Nematodes also serve as important bio-indicators of environmental health in environmental studies, offering insights into soil quality and ecological balance. These organisms can be found in virtually all environments around the world. The term nematode has its roots in Greek, deriving from the words "nema" and "oides", which mean thread and resembling, respectively. Nematodes are an amazingly diverse group of organisms that can either be beneficial or parasitic to plants and animals alike. Generally, they have a slender body that is transparent and lacks segments, exhibiting bilateral symmetry. The study of nematology has made significant strides in the field of medicine. Notably, it has played a pivotal role in developing new antibiotics for the treatment of bacterial infections. Furthermore, nematology has been instrumental in identifying that tropical diseases like elephantiasis and ascariasis are caused by a type of nematode. Ascariasis is a parasitic infection caused by *Ascaris*. The disease occurs when individuals ingest food or water contaminated with the eggs of *Ascaris* species, typically found in soil, vegetables, fruits, and other foods. Research has revealed that *Ascaris* infections can potentially lead to impaired cognitive function in certain school-aged children. This underscores the importance of proper hygiene and food safety measures in preventing the spread of this disease. Several nematodes cause diseases in animals, such as fish, where they reduce the quality of the meat and pose a risk to food security. Nematodes are a type of parasite that can infect various types of fish in freshwater, marine, and brackish water environments. Some nematodes can have devastating effects on wild fish populations, leading to significant fish mortality. These parasites can infect fish in their adult stage, but their larval forms can also infect fish species after passing through birds, mammals, or reptiles that consume fish, or even through predatory fish. Certain nematodes, like *Anisakis*, are zoonotic, meaning they can be transmitted to humans. Consumption of raw/undercooked infected fish meat poses a risk of infection to humans. Nematodes also pose a threat to the economic value of fish due to consumer concerns about their presence in food products. Infected fillets are often rejected, leading to increased production costs. Therefore, research on nematology brings valuable insight into the fish/animal parasites aiming to secure food.

In agricultural research, nematology plays a crucial role in providing a wide range of ecosystem services that have a significant impact on the nitrogen cycle, the ability of soil to decompose waste, and the control of pests within soil systems. Although, certain female plant-parasitic nematodes can take on a spherical shape, such as cyst and root-knot females. All plant-parasitic nematodes possess a stylet in their anterior end that helps them to pass the food through the plant cells to their body. This feature enables plant-parasitic nematodes to inject the enzymes into plant cells, which digest the food and help nematodes to develop and cause damage to plant cells. Plant-parasitic nematodes pose a significant threat to various plant types, including vegetables, trees, turfgrass, and foliage plants. They can cause extensive damage and significantly reduce crop yield. Root-knot, cyst, root lesion, spiral, burrowing, bulb and stem reniform, dagger, bud and leaf, and pine wilt disease are among the most harmful nematodes. These pests are responsible for an average loss of 12.3% annually in 40 major crops worldwide, with developing countries bearing the brunt of the losses, estimated at 14.6%, compared to 8.8% in developed nations. Plant-parasitic nematodes result in global economic losses of nearly $125 billion per year, affecting all agricultural crops. The impact of harmful nematodes on agricultural production is regularly undervalued, as their symptoms are frequently mistaken for other issues such as water stress, nutritional disorders, virus infection, soil fertility problems, or complex diseases caused by interactions of fungal/bacterial with nematodes. The severity of their impact depends mostly on the population density in the soil and roots, the cultivar susceptibility, and the ecological circumstances. Root-knot and cyst nematodes are the main destructive plant-parasitic nematodes. The implications of these misinterpretations can be significant, as they can lead to the implementation of inappropriate corrective measures, resulting in further damage to crops and reduced yields. Therefore, an accurate diagnosis of nematode infestation is critical in ensuring optimal agricultural productivity and profitability. To achieve this, it is essential to utilize appropriate diagnostic techniques to differentiate nematode symptoms from those of other conditions. This approach can help farmers and agricultural experts make informed decisions regarding the management and treatment of nematode-infested crops, ultimately leading to better outcomes for all involved.

On the other hand, beneficial or free-living nematodes play a critical role in soil health due to their contribution to soil nutrition, nitrogen fixation, and microbial balance. The soils in a hectare of all agroecosystems typically contain billions of both plant-parasitic and beneficial nematodes, which can significantly affect crop yields. In conclusion, nematology discipline is an attractive and principal field of study that plays a vital role in understanding the complex relationships between nematodes, plants, humans, animals, and their environment.

CHAPTER 2

Sampling

Abstract: The process of sampling nematodes has become a crucial aspect of agricultural research. The accurate identification of these pests is essential, and the method of sampling is dependent on whether they are plant-parasitic or free-living. To ensure that errors are minimized and samples are reliable and representative, the sampling pattern is based on the area being studied. This can range from randomized to systematic sampling techniques. Furthermore, the final results are influenced by various factors such as the timing of sampling, depth of samples, and the total number of samples taken.

Keywords: Depth, Pattern, Sampling tool, Soil sampling.

GENERAL CONSIDERATION

Useful nematode sampling depends on the time and the target group of the nematodes, which can differ. Several factors, including ecological consideration and distribution patterns, should be considered for plant-parasitic nematodes [1, 2].

As nematodes cannot scurry off within the soil, their damage to crops related to symptom expression on aerial plant parts appears as spots of general low growth (*e.g.*, stunted and/or wilted, yellowish plants), which fluctuates throughout the growing season. The best time for the sampling is when the soil is damp, not too wet or too dry. Nematodes are present in soil rhizosphere samples during the entire growing season, although their population densities fluctuate according to various abiotic and biotic factors. Although nematodes may exist in the soil in all seasons, the larval stages of plant-parasitic nematodes are more apparent in the winter samples. Free-living nematodes of the family Cephalobidae and many belonging to the family Rhabditidae appear more frequently in the samples during autumn.

On the other hand, plant-parasitic nematodes have high population densities during the growing season when climatic conditions and plant growth are optimal.

Bacterivores nematodes belonging to the Rhabditida family mainly exist in substrates with high organic material contents such as manure or dung. Predator nematodes of the order Mononchida mostly live in the wetlands or the borders of aquatic areas (*e.g.*, rivers, lakes, lagoons, etc.). All nematode groups generally occur in the larval stages during winter. Concerning plant-parasitic nematodes, in particular, sampling should be done both around patches where plants appear to be growing optimally and where plants are stunted, wilted, or yellowish [3]. This will ensure a proper comparison of the nematode status of 'healthy' versus 'infected' plants.

TOOLS FOR SAMPLING

Augers, Edelman, and Helical (Fig. **1**) are some of the tools that can be used to sample nematodes at a depth of 30 cm or more [4]. However, in dry soil, it is difficult to insert an auger properly for accurate sampling. Garden trowels, narrow-bladed shovels, big kitchen spoons, or spades are also useful tools to sample nematodes, mostly when sampling is done in areas with rocky soil (Fig. **1**). All equipment should be cleaned or appropriately sanitized, *e.g.*, washed with or soaked in a 1% NaOCl solution, after each sampling activity to avoid contaminating nematode populations that occur in different areas where sampling is done for separate projects.

A proper auger is essential for taking soil samples from various depths associated with multiple crops (Fig. **2**).

SAMPLING FROM AQUATIC AND WETLAND

Nematodes are present in benthos belonging to various groups and families. However, they cannot be collected from enough individuals. Therefore, constant and hard work is required to assess the proper number of nematodes. The device for sampling from benthos is shown in Fig. (**3**).

SIZE OF THE SAMPLING AREA

The size of the sampling area must be adapted to fit the purpose of the study. The sampling location should typically be divided into smaller pieces, for example, one hectare. Generally, a nematode sample (soil, roots/other below-ground plant parts and/or aerial plant parts) should consist of several sub-samples (at least 10) of each area being sampled. About 20 sub-samples can be combined to form one mixed sample of 1-2 kg for 5 acres [5]. For example, mixing soil can be done by hand to avoid mechanical damage to the nematodes. A sub-sample may consist of at least 100 g of soil and 5-50 g of plant material. Plant material should be chopped into small pieces, *e.g.*, 1cm, and appropriately mixed to obtain a representative sub-sample. In or near aquatic areas, the sample that is in a solution form should be transferred to a plastic container while the soil is kept in a plastic bag (Fig. **4**).

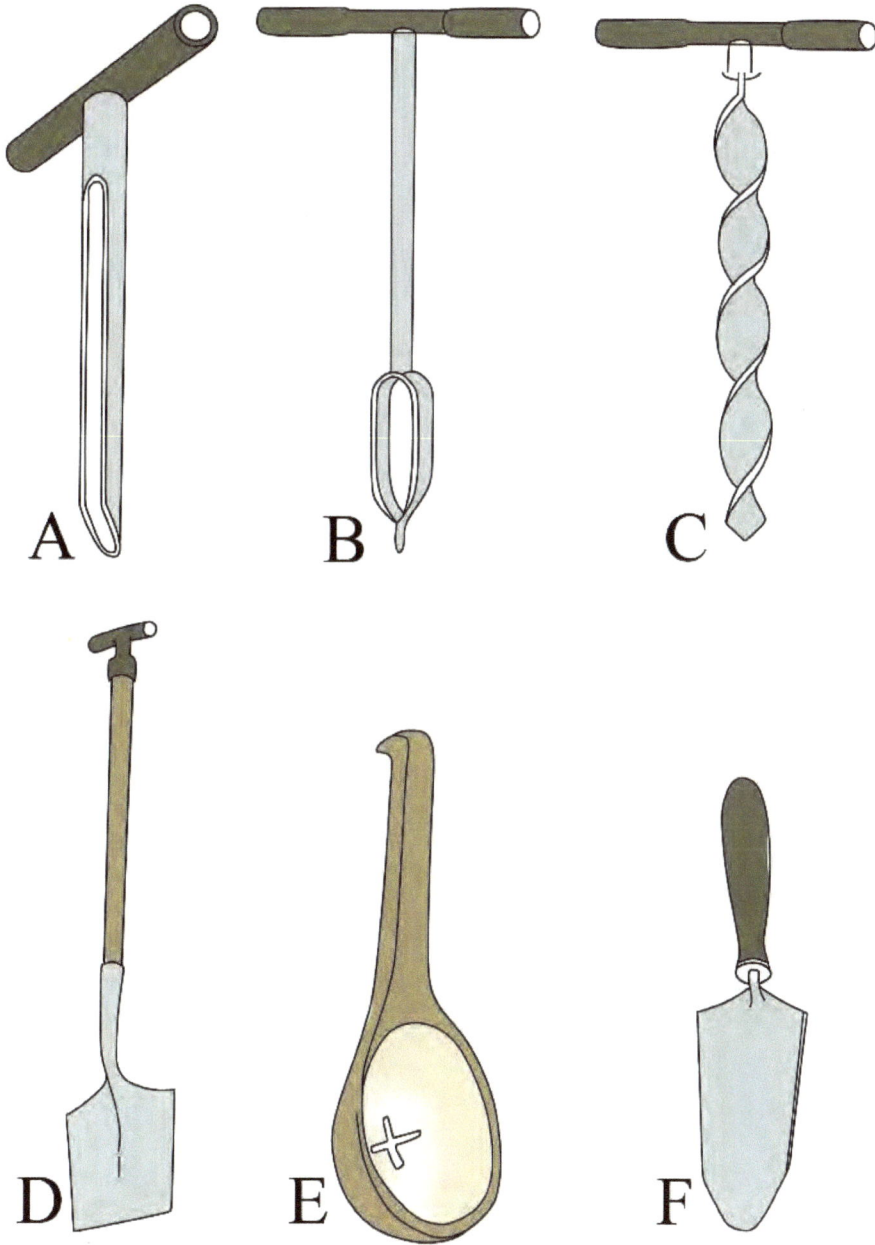

Fig. (1). Tools generally used for nematode sampling (adapted from Kleynhans [4]).

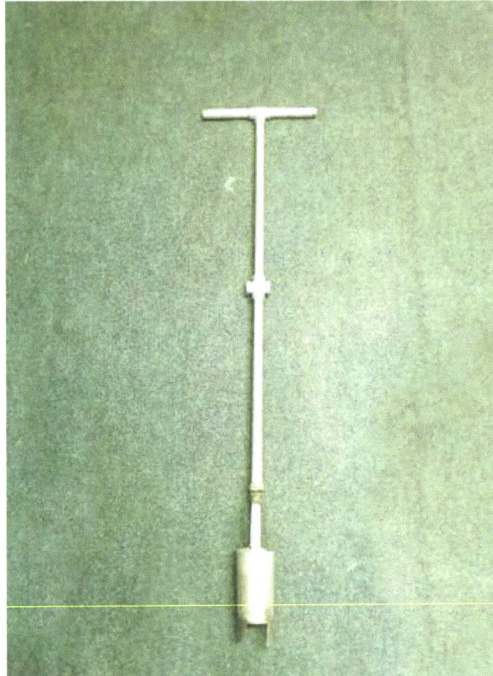

Fig. (2). An auger for soil sampling (designed and built by Mr. Geldenhuys, Aquaculture Research Unit, University of Limpopo, South Africa, 2022).

Fig. (3). A device for benthos sampling with a 25 μm sieve at the base (designed and built by Mr. Geldenhuys, Aquaculture Research Unit, University of Limpopo, South Africa, 2022).

Fig. (4).   Water-based samples can be sampled in plastic containers (A), while plastic bags (B-F) can be used to keep soil and plant tissue for nematode extraction.

For biodiversity purposes, 500 g to one kg of soil must be collected. A soil sample of 160 g (representing a combined sample from various sub-samples) per sampling site is adequate for chemical and physical analysis, while 100-200 g [6, 7] or more [8] is adequate for extracting and counting nematode individuals. For taxonomical purposes, such as description of new species, at least 500 gr of soil is needed to ensure that more specimens can be obtained for morphological and morphometrical analyses if necessary.

SAMPLING DEPTH

Depending on root growth, soil texture, and nematode species, nematodes can be found up to a few meters into the soil profile. Generally, most of the nematode population can be found in the upper 0 - 30 cm of the soil, which usually represents that part of the soil profile where the bulk of plant roots/other below-plant parts are located. In orchards, sampling is done at greater depths since the roots of trees stretch more in-depth into the soil than those of annual crop plants. Generally, nematode individuals of the families Cephalobidae and Rhabditidae, plant-parasitic nematodes of the suborder Tylenchina, predator nematodes of the order Mononchida, and most of the freshwater nematodes occur in the upper 0-30 cm of the soil. However, the rhabditid species *Halicephalobus mephisto* has been detected 0.9-3.6 km deep in water in a deep mine in South Africa [9]. In another survey, a member of the Cephalobidae family was recovered from the soil at a depth of 1 m in a mountain [10]. Some nematodes, such as root-knot (*Meloidogyne* spp.), pin (*Paratylenchus hamatus*), dagger (*Xiphinema index*), and needle (*Longidorus* spp.), may also infect deep-rooted hosts and hence require sampling at soil depths of up to 45-90 cm [11].

SAMPLING PATTERN

The purpose of the study is to depict the sampling pattern [2]. As plant-parasitic nematodes are not distributed equally in fields, samples should be collected from several areas across the field to ensure that they represent the nematode distribution pattern. For taxonomical studies, samples can be collected randomly across a designated area (Fig. **5**). For diagnostic purposes, samples should be collected from low-growth areas and areas where plants are growing optimally to enable comparison (Fig. **6**). For biodiversity purposes, the field should be divided into 1-hectare areas, and samples should be collected systematically across the whole site (Fig. **5**). For annual crops planted in rows, 10-30 cm rhizosphere soil samples from around the stem should be managed. For trees, rhizosphere soil and feeder roots should be sampled at least 4-10 places around the trunk (Fig. **6**).

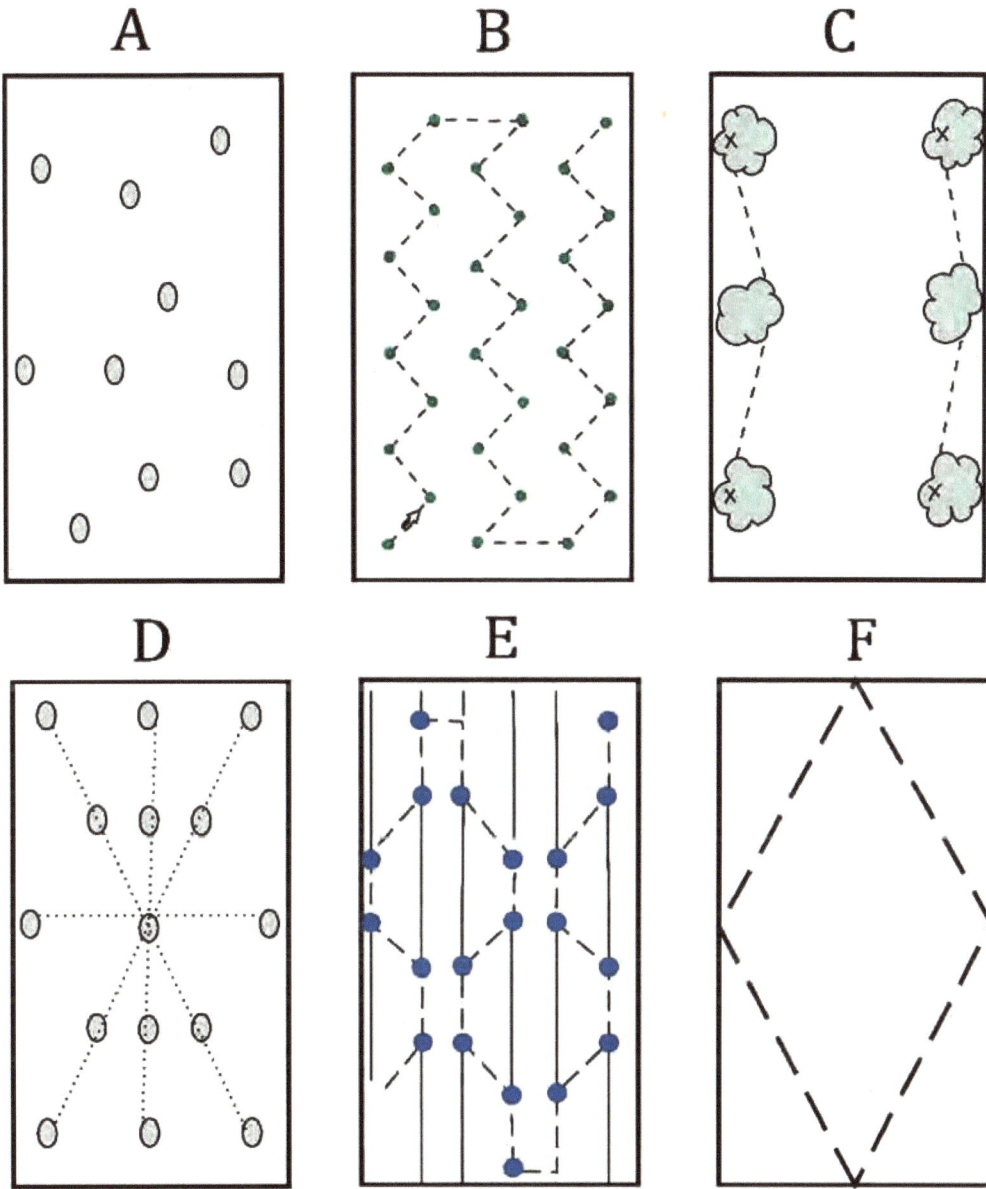

Fig. (5). Different patterns used to ensure accurate nematode sampling in annual crops (adapted from Santo *et al.* [5]).

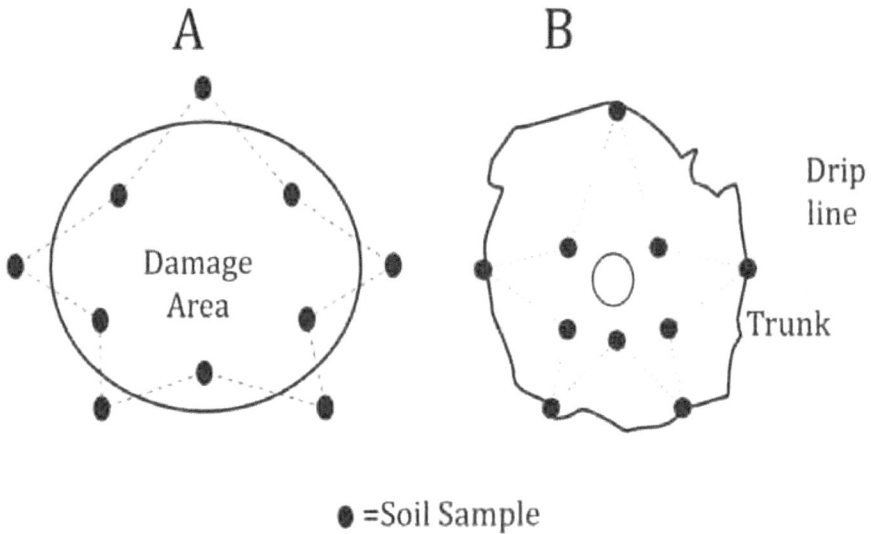

Fig. (6). Patterns of sampling used around trees to enable the establishment of nematode population levels and/or diversity (adapted from Kleynhans [4]).

The label of samples should contain information about 1) the crop and cultivar; 2) the sampling date; 3) Global Positioning System (GPS) coordinates of the locality and, if possible, a photo of the host plant or area; 4) the history of the field in terms of cropping sequences (at least for 3 – 6 years), and 5) information about the next crop to be planted.

Soil and plant samples should be transferred to a nematology laboratory quickly or kept in a cool box (4 °C) during transportation to the laboratory. If a cool box is not available, the samples should be kept out of direct sunlight. Water should never be added to samples (*e.g.*, if the soil was dry during sampling) to avoid any changes in the natural composition of the nematode communities and other microorganisms (that may impact the nematode communities) that are contained within such a sample. Samples can be stored at 4-6 °C for up to six months for taxonomical studies and about three months for biodiversity and diagnostic purposes without considerable harmful effects on nematode numbers to be expected. However, for trustworthy analyses of nematode population densities and diversity, samples should be processed within 5-7 days after sampling and storage at 4-6 °C. As bacterivorous nematodes (*e.g.*, Rhabditida) multiply even under low temperatures, their numbers and species composition will change during storage, with the same being experienced for cryophilic plant-parasitic nematode species. Should the processing of nematode samples be expected to be delayed, fixing the nematodes present in such samples using formalin (4-10%) is recommended to preserve the samples until they can be transferred to the laboratory [12 - 14]. For wet samples, mainly sediments, formalin concentration should be adjusted to ensure the optimal concentration is used for proper fixing of nematode individuals.

CONCLUSION

The process of soil sampling is an essential and meticulous step in the study of nematodes, where the accuracy and precision of the collected samples can significantly impact the quality of the results. The importance of soil sampling, however, depends on the research objectives, which can range from understanding the diversity of nematode species in a particular habitat to investigating the ecological and molecular mechanisms underlying their behavior and interactions with other organisms. A rich source of nematode individuals obtained through soil sampling can provide researchers with a valuable foundation for further analysis of various subjects in nematological research, such as taxonomy, population dynamics, biogeography, and evolution, to name a few. Therefore, the quality and quantity of nematodes collected through soil sampling can open doors to new avenues of exploration and discovery in the field of nematology.

CHAPTER 3

Nematode Extraction

Abstract: Obtaining an adequate number of nematode specimens is crucial for various studies related to biodiversity, taxonomy, and management. The extraction method chosen for the nematodes depends on the specific study being conducted. In regards to sedentary endoparasites, the roots are the primary source for nematode extraction, while for other purposes, the soil is the preferred option. It is essential to evaluate the efficiency of the extraction method used. Nematode extraction must be performed promptly after sample collection since the specimens tend to deteriorate with time. In this chapter, some of the commonly used techniques for nematode extraction are given.

Keywords: Efficiency, Extraction, Root, Soil.

EXTRACTION OF SOIL AND ROOT NEMATODES

Tray Method

This is the simplest and cheapest method that is adapted from the Baermann Funnel technique. Even slow-moving nematodes, such as Criconematids, can be extracted using this technique. This method is relatively fast for yielding large numbers of alive and active nematodes and is suitable for ecological and taxonomical studies of free-living and plant-parasitic nematodes [10].

Procedure

1. Put on gloves for safety (Fig. **1**).

2. Weigh 200 g of soil for biodiversity investigation and up to 500 g for a classical study.

3. Mix the soil.

4. Spread the soil on tissue or towel paper supported on a coarse meshed plastic screen kept in a plastic container.

5. Add tap water to cover the soil surface.

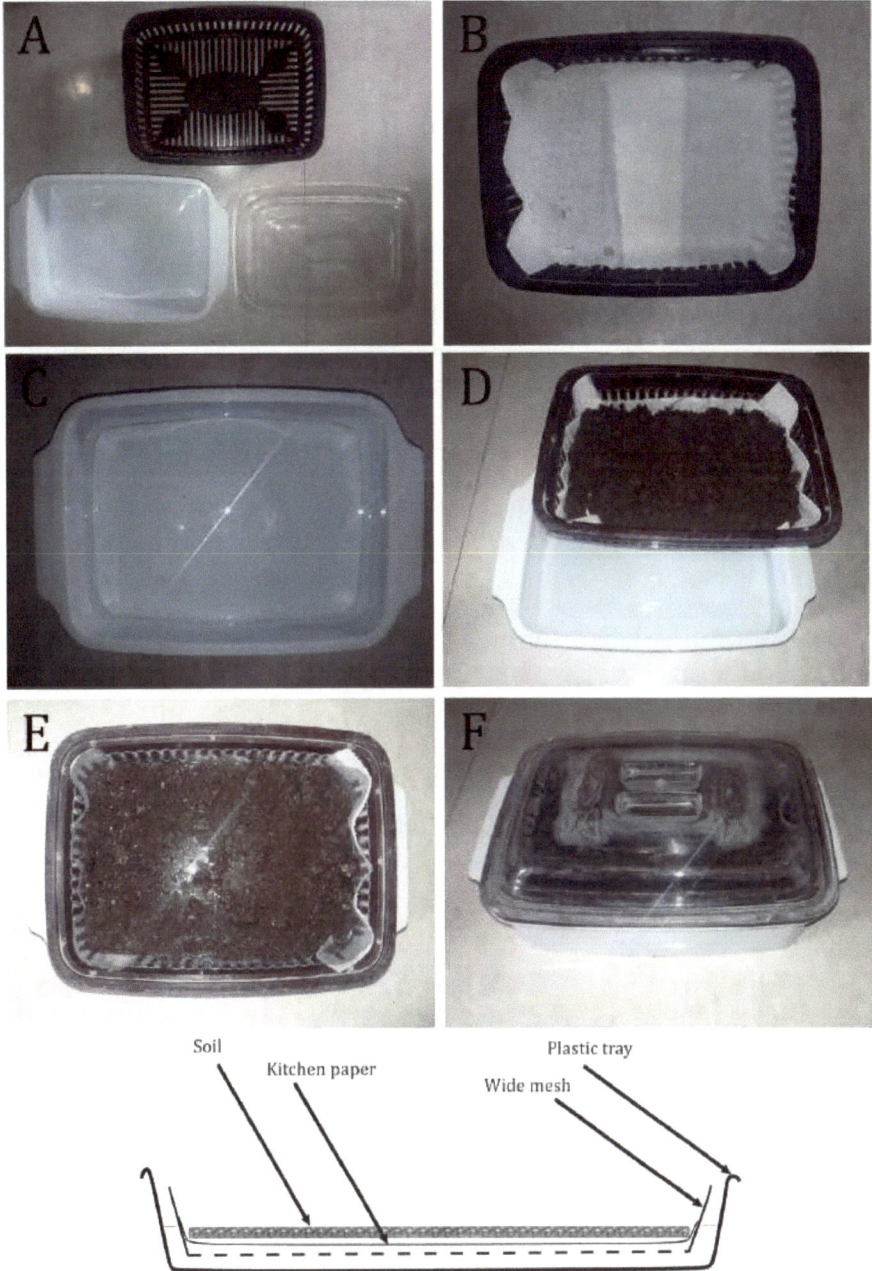

Fig. (1). The tray method that is used to extract nematodes from soil using a tray, including (**A**) three plastic parts; (**B**) a plastic mesh part covered with tissue- or towel paper; (**C**) the tray filled with an appropriate amount of water; (**D**) soil over the tissue paper; (**E, F**) soil over the tissue paper covered by a thin layer of water; (down illustration) schematic plan of entire tray method.

6. Leave the soil to soak in water in the container for 18-24 hours or at least overnight.

7. Cover the container to avoid evaporation of the water.

8. Collect the solution at the bottom of the container into Petri dishes.

Sugar Flotation Method

This method is fast and yields a high number of nematode individuals [2, 10]. However, more equipment and materials are needed than the tray method. By using this method, sluggish and active nematodes are extracted successfully. This method is very suitable for studies on plant-parasitic nematodes for ecological and classical purposes.

Procedure

1. Mix the soil with water and pass the solution through a 60-mesh size sieve to remove the debris material.

2. Next, pass the soil solution through the 100, 120, and 400 mesh size sieves, stacked from bottom to top.

3. Collect the residue on the 400-mesh size sieve and transfer it to a 500-1000 beaker or graduated cylinder and leave it for 20-30 minutes for dust, hummus, and debris to move up to the water surface.

4. Remove the debris from the water surface.

5. Shake the solution in the graduated cylinder and pour 300-500 ml of that into centrifuge tubes.

6. Centrifuge the solution for 5 min at 5000 rpm.

7. Take the tubes out and remove the water.

8. Prepare the sugar solution as follows: Put 700 ml of sugar into a 1000 ml graduated cylinder, then add water up to a volume of 1000 ml (hot water is preferable); shake the cylinder to dissolve the sugar in the water.

9. Add the sugar solution to the centrifuge tubes, then mix it with the soil that had settled at the bottom of the tubes.

10. Centrifuge for 1 min at 1000 rpm.

11. Quickly take the tubes out and pour the solution onto the 400-mesh size sieve and simultaneously add water to the solution to avoid denaturing the nematodes.

12. Pour the residue that was collected on the 400-mesh size sieve into a petri dish.

13. Observe the nematodes under the stereomicroscope.

Incubation Method

This method is easy to conduct; however, the nematodes extracted may not be in good condition because of the low oxygen concentration. This method is suitable for extracting the nematodes from leaves, kernels, roots, or tubers. This method can yield high numbers of *Robustodorus* and *Ditylenchus* from groundnut kernels and hull material incubated for 24 h at 25 °C [15].

Procedure

1. Cut the plant material infected with nematodes into small pieces (Fig. **2**).

2. Put the pieces in a petri dish or another container.

3. Add tap water until the material is covered and leave it for at least 24 hours.

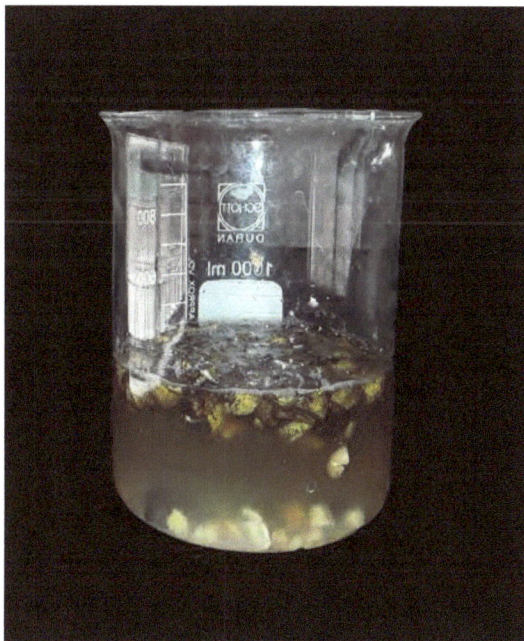

Fig. (2). The incubation method used for aphelenchoid extraction (photo by Ebrahim Shokoohi).

Sieving Method

This technique provides quick results; however, tiny nematodes may be lost during processing.

Procedure

1. Wash and decant 50 ml of soil, and pass it through a 60-mesh size sieve to remove the debris.

2. Pass the solution through a 45 μm sieve.

3. Collect the residue in a 200 ml graduated cylinder.

4. Add five drops of a flocculent (*e.g.*, kaolin) to the cylinder.

5. Close the cylinder and shake it several times.

6. Allow the suspension to settle for 20 second.

7. Pour the suspension into a 200 ml cylinder.

8. Add five drops of a flocculent to the cylinder and mix by shaking it several times.

9. Leave the suspension for 10 second.

10. Pour the supernatant into a 200 ml cylinder.

11. After 50 min, remove most of the suspension and leave 20 ml for counting nematodes present.

Dissecting Method

This method removes any life stages, especially females of root-knot, citrus nematodes, and other sedentary plant-parasitic nematodes that feed inside plant parts, using a scalpel or needle and fine forceps. Besides, the egg mass of these nematode genera can also be isolated. If the roots are stained, the nematodes are more clearly visible and more comfortable to locate and isolate. This method is handy for molecular studies on *Meloidogyne* when females are selected for DNA extraction. Furthermore, females are also easily obtained using this method for morphological studies, *e.g.*, analyses of perineal pattern morphology.

Seinhorst Cyst Extraction Elutriator Method

This method is used to extract cysts, *e.g.*, *Heterodera* and *Globodera*, from the soil. Cyst nematodes can be extracted from dry and wet soils. This method is expensive, and a large amount of water is needed for extractions.

Procedure

For optimal results when using the elutriator, please follow these instructions (Fig. **3**).

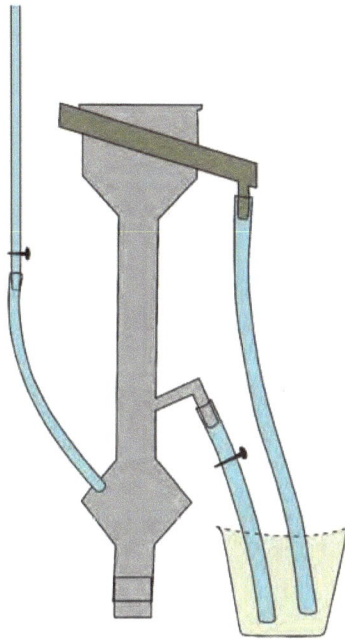

Fig. (3). A schematic plan of the Seinhorst tool for cyst extraction.

First, place a 200 or 250 µm sieve underneath the overflow collar and side outlet. Then, fill the elutriator with tap water using a water stream of 3500 ml per minute. Next, filter your sample through a 2 mm sieve into a funnel and thoroughly clean it by moving the sieve up and down. The waiting period will vary based on soil type, generally lasting between 2-5 minutes until the water overflow runs clear. Afterward, turn off the upward water current and rinse the outlet collar to ensure all cysts are washed onto the 200 or 250 µm sieve. Then, open the side outlet to let the suspension containing heavier cysts pass through the sieve. You can retrieve the cysts from the sieve for direct examination or further cleaning. Once finished with the elutriator, open the bottom outlet and rinse it with water. Finally, count the cysts and use a dissecting microscope to identify them.

Baunacke Method

Baunacke [17, 18] presented a method for the collection of *Globodera* and *Heterodera* cysts from the soil. During this method, dried cysts that float in water are collected on sieves. This method is cheap and fast; however, the success rate depends on the technical assistant's expertise.

Procedure

1. Pass the dried soil through a 200 or 250 μm sieve with a jet of water to remove debris and extra soil particles (Fig. **4**).

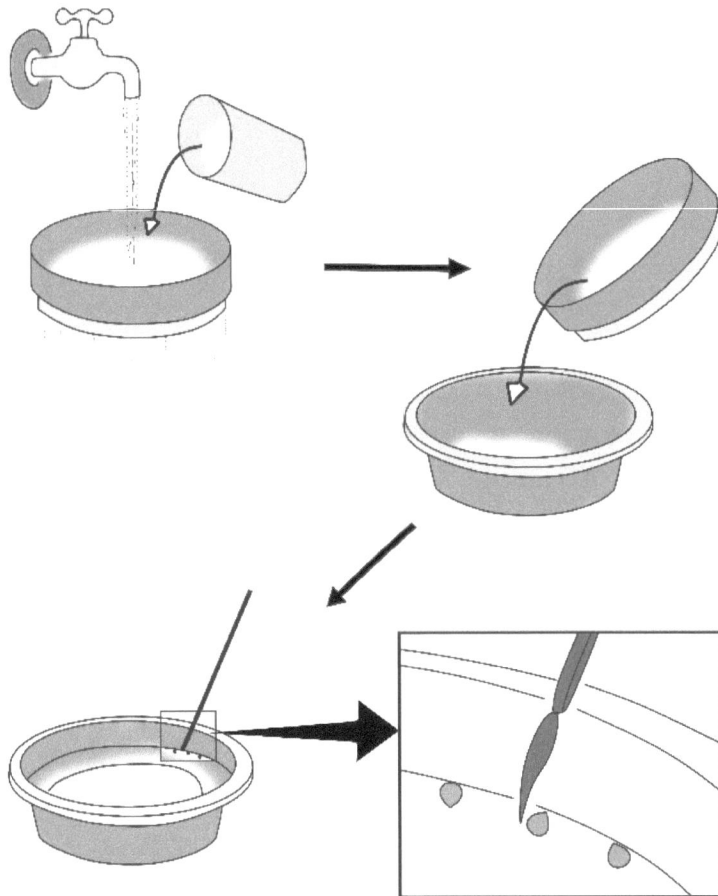

Fig. (4). Schematic view of Baunacke method of cyst extraction using a 180 μm sieve (a 200 to 250 μm sieve for *Globodera* and *Heterodera glycines* is suitable; Adapted from EPPO [17]).

2. Transfer the debris remaining on the 200 or 250 µm sieve to a plastic beaker or bowl.

3. Stir the suspension thoroughly.

4. Let the suspension settle for at least 30 s.

5. Add a drop of detergent that will encourage the cysts to move to the edge.

6. Transfer the cysts to a dish using forceps or a fine painting brush and inspect under a dissecting microscope.

Maceration and Filtration or Blender Method

This method is suitable for extracting motile and immotile stages of *Hirschmanniella* spp., *Radophulus similis*, *Nacobbus aberrans*, and other endoparasitic nematodes from roots, tubers, and other plant tissue. Further details on this method are given by Coolen *et al.* [19] and Seinhorst [20]. This method is also suitable for extracting eggs and juveniles of sedentary nematodes, such as *Meloidogyne* and *Tylenchulus*, from roots/other plant tissue. This method is easy to do; however, the nematodes' adult stages may be damaged during the extraction.

Procedure

1. Cut the roots into small pieces.

2. Put the root pieces into a kitchen blender and add 100-200 ml tap water.

3. Blend for 1 min on high speed.

4. Pour the solution through a 120-mesh size sieve to remove the debris, then through a 400-mesh size sieve.

5. Collect the residue on the 400-mesh size sieve and transfer it to Petri dishes for counting/identification.

Wood and Compost Extraction

Rotten woods and composts are the source of various groups of nematodes, including aphelenchoids and rhabditids. Additionally, spongy tissues, such as fungi at the base of rotten woods, inhabit free-living nematodes such as *Panagrolaimus* [21]. For extraction from wood parts, they need to be a grinder, and then, nematodes can be extracted through the Baermann funnel or tray method.

EXTRACTION OF ENTOMOPATHOGENIC NEMATODE (EPN)

Nematode Isolation

Entomopathogenic nematodes (EPNs), including *Steinernema* and *Heterorhabditis,* are soil-dealing nematodes. These nematodes' isolation is different than plant-parasitic and free-living ones. Therefore, the last instar larvae of *Galleria mellonella* (Lepidoptera, Galleriidae) are used to isolate the EPN from soil habitats. *Galleria mellonella* is placed into a 50 ml plastic container [22] at room temperature (about 22-25°C) or 18°C in a dark condition. Soil samples are then checked every 2 days from the 4th day after the insertion of Galleria larvae. Dead insects with reddish or red-violet color are collected from the soil samples, rinsed in distilled water, and placed on White [23] traps to collect emerging IJ [24].

Practical Procedure

The procedure includes soil sampling, isolation of EPN from the Galleria trap, and recovering nematode from the White trap (Fig. **5**) [24, 25].

1. Collect the soil samples at a depth of 30 cm (5 random samples per area).

2. Transfer the soil samples with proper information (GPS coordinate information, date, vegetation, etc.) in a cool box to the laboratory.

3. Place approximately 200 to 250 ml of moist soil in a clean plastic container with a lid.

4. Add Galleria baits, at least four; however, 5-10 last instar Galleria larvae can be suitable.

5. Cover the container with a lid and turn the container upside down.

6. Maintain containers in the dark and at room temperature (usually 22 to 25 °C).

7. Check containers every 2-3 days and remove dead insects. By four days, if there is EPN in the soil, they will kill the insects. Additionally, new larvae of Galleria can be added to the container of the soil sample.

8. Remove dead insects from the containers. Cadavers with a brown or ochre coloration are usually parasitized by steinernematids, whereas brick red to dark purple cadavers are parasitized by heterorhabditids.

9. Rinse cadavers in sterile water.

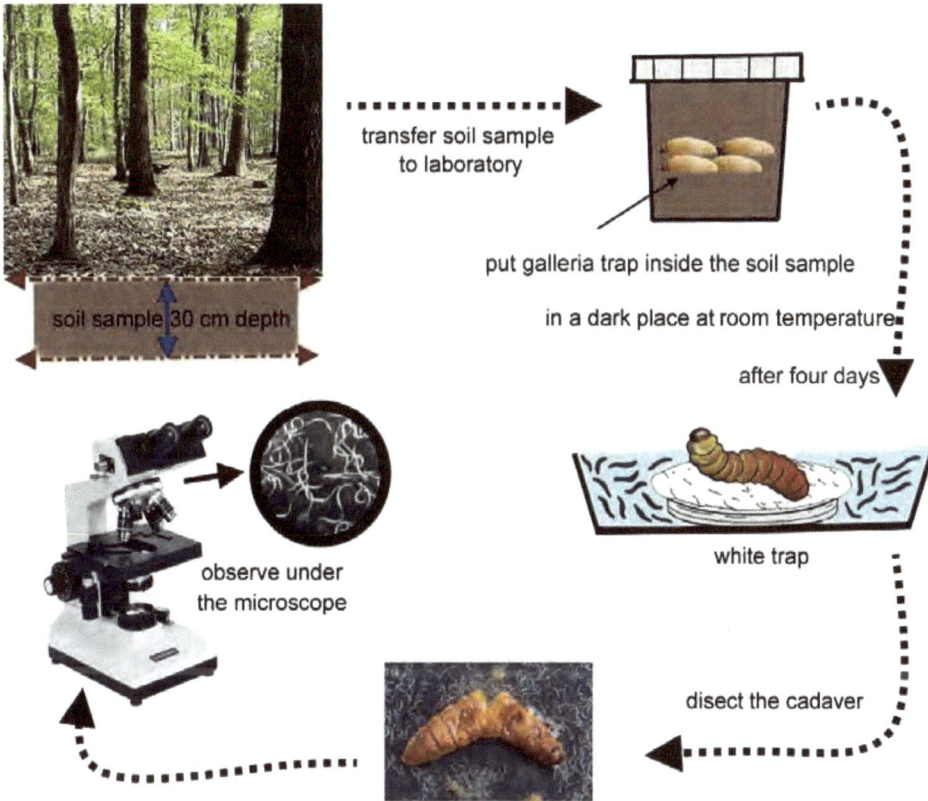

Fig. (5). Isolation of EPN from the soil using Galleria and White trap method.

10. Place cadavers in a White trap for recovery of nematode progeny.

11. Place the top of a 50 (or 60) mm diameter Petri dish inside a larger container (100 mm).

12. Set one single circular filter paper (Whatman #1) inside the smaller dish.

13. Place cadavers on the filter paper of the smaller dish, making sure they do not touch each other to avoid any contamination.

14. Fill the larger Petri dish with ca. 20 ml of sterile distilled water. Do not add water to the dish that holds the cadavers.

15. Cover the large Petri dish and its contents with the lid. Note all necessary information, such as infection date, trap date, and nematode species.

16. Keep trap at room temperature until the emergence of juvenile infective stages (IJs) occurs. This process can take between 10-25 days, depending on the nematode species or strain considered.

17. Remove water with IJs from the larger dish of the trap and pour water with nematodes into a beaker.

18. Allow nematodes to decant to the bottom of the beaker. This process may take a few minutes.

19. Pour water carefully, ensuring nematodes remain in the bottom of the beaker.

20. Rinse nematodes by adding more water and allowing nematodes to decant. This step can be repeated 2-3 times until the water is clean.

21. Place nematode suspension in a tissue culture flask (250 ml). Keep the concentration of the suspension to 1000-3000 nematodes/ml.

22. Store the flask with nematode suspension in a cold room or in an incubator between 10-20 °C. Check stored flasks periodically, as the shelf life of EPN is variable. Usually, steinernematids can be stored for 6-12 months without the need for subculturing, whereas heterorhabditids may require more periodic check-ups.

Investigation on Parasitic Nematode of Mosquito

Mosquitoes are capable of transmitting serious diseases such as chikungunya, dengue, yellow fever, and malaria. Among these diseases, malaria stands out as the deadliest disease worldwide. Mosquitoes, including *Anopheles,* are important in the medical field. Several authors indicated infection of mosquitoes with mermithid nematodes, such as *Romanomermis* species [26 - 29]. The second stage of *Romanomermis* is to locate the host accidentally and then enter the insect through the cuticle. Afterward, the nematode goes to the hemocoel [26]. Therefore, the infection of the mosquitoes by the nematodes opens a door to biological control of the dangerous mosquitoes that threaten human health.

Practical Procedure

Mosquitoes Preparation in the Laboratory

1. Mosquitoes, including *Culex pipiens* and *Anopheles,* can be collected from the infected areas.

2. Transfer the mosquitoes to the laboratory with the following conditions: 28 ± 2 °C, a relative humidity of 70-80%, and a photocycle of 12 hours dark and 12 hours light.

3. Rear the mosquitoes in a bucket containing dechlorinated tap water.

4. Feed the larvae using a 1:3 mixture of yeast and ground wheat rusk daily.

5. Pick up the pupae daily and transfer them into a glass with dechlorinated tap water and then transfer them to a labeled rearing cage for adult emergence.

6. Feed the adult stages using a cotton pad soaked in a 10% sugar solution. Female mosquitoes can be fed using the blood meals of pigeons.

Propagation of Mermithid Nematodes

1. Prepare a second larvae of *Culex quinquefasciatus* as a host for *Romanomermis* species.

2. Prepare a chicken to supply these mosquitoes with blood meals.

3. Use a small plastic tray containing about 500 ml of water in each mosquito cage for oviposition.

4. Four days after a blood meal, collect the eggs (about 6 eggs) and deposit them in a plastic tray containing 2 litres of water.

5. Use several plastic trays in a room with the following conditions: room temperature and relative humidity of 28 ± 2 °C and 70–90%, respectively.

6. Cover the containers with a mesh sieve to avoid oviposition by wild mosquitoes.

7. The mosquitoes of second-stages larvae are infected by *Romanomermis* nematodes.

8. After 8 weeks, pick up the nematodes. Then, flood the cultures with sterilized water to induce the exclusion of eggs and the emergence of infective pre-parasites from the substrate.

9. Then, fourteen hours after flooding the cultures, decant the water and calculate the nematode concentration.

10. After 8 days, the mosquito larva will die and appear on the water surface, indicating the end of the parasitic phase of the nematode.

11. Then pour the water on the sieve containing post-parasitic juveniles (J4) together with the dead mosquito larva and other debris.

12. Place the sieve containing the larvae and J4 in a container with clean water.

13. After a few minutes, the J4 passes through the sieve into the clean water so that when the sieve is removed, the dead mosquito larva and debris will be separated from the J4 that settled on the bottom of the container.

14. Collect the J4 using a syringe and transfer it to a glass beaker with clean water. Then wash the J4.

15. Deposit about 2 grams (it may vary depending on the experiment) of washed J4 in round plastic containers with previously sterilized coconut coir fibers (about 35 g) and 500 ml sterilized (chlorine-free) water.

16. After about 3 hours (when all nematodes had moved into the substrate), decant the water and cover the containers for eight weeks so that the nematodes can reach sexual maturity, mate, and deposit eggs.

It should be noted that condensation water droplets should be removed regularly (each week) from the containers with cotton tissue to prevent premature hatching of nematode eggs.

Evaluation of Different Extraction Methods

Various techniques for nematode extraction have advantages and disadvantages (Table **1**). Additionally, each method requires a specific device, which may affect the cost of nematode extraction. Therefore, selecting the proper technique based on the need of the study is critical.It allows the extraction of all stages of nematodes, possibly with 100% efficiency [30]. For example, centrifugal flotation methods are only suitable for isolating slow and inactive nematodes [31]. On the other hand, the tray method separates the active and motile stages of the nematodes, but not sluggish ones such as Criconematidae, unless they have a high number in the tested sample [32]. In conclusion, temperature and water usage also affect the method to be used, which affects the extraction efficacy and quality.

Table 1. Evaluation of different extraction methods for soil and root nematodes [17].

Extraction Method	Principle	Maximum Sample Size	Extraction Efficacy	Cost of Equipment	Labor Costs	Water Use	Time Until Evaluation*	Quality of Extraction
Plant material								
Direct examination	Motility	10 g	+	+	+	+	10 min	+

(Table 1) cont.....

Extraction Method	Principle	Maximum Sample Size	Extraction Efficacy	Cost of Equipment	Labor Costs	Water Use	Time Until Evaluation*	Quality of Extraction
Baermann funnel/Oostenbrink	Motility	50 g	++	+	++	+	24 h	+++
dish	-	-	-	-	-	-	-	-
Root incubation	Motility	20 g	++	+	++	+	72 h	++
Mistifier	Motility	50 g	+++	++	+	+++	24 h	++
Maceration and filtration	Size and shape	50 g	+++	++	++	++	15 min	+
Maceration and centrifugal	Density	50 g	+++	+++	++	++	30 min	++
flotation	-	-	-	-	-	-	-	-
Enzymatic digestion	Size and shape	10 g	++	++	++	+	72 h	+
Soil								
Baermann funnel/Oostenbrink	Motility	250 mL	+	+	++	+	24 h	+++
dish	-	-	-	-	-	-	-	-
Flotation and sieving	Density and	200 mL	++	++	+++	++	15 min	+
	size and shape	-	-	-	-	-	-	-
Flegg modified Cobb	Density and	1000 mL	++	++	+++	+++	24 h	++
Oostenbrink elutriator and	size and shape Density	250 mL†	+++	+++	++	++	24 h	+++
Baermann funnel Oostenbrink elutriator and	Density	250 mL†	+++	+++	++	++	60 min	+++
centrifugation	-	-	-	-	-	-	-	-
Centrifugal flotation	Density	250 mL	++	+++	++	++	15 min	+++
Cysts								
Baunacke method	Density and	100 mL	+	+	++	+	10 min	+
Paper strip method	size and shape	-	-	-	-	-	-	-
Fenwick can	Density	250 mL	++	++	++	++	15 min	++
Schuiling centrifuge	Density	500 mL	++	+++	++	+++	15 min	++

(Table 1) cont.....

Extraction Method	Principle	Maximum Sample Size	Extraction Efficacy	Cost of Equipment	Labor Costs	Water Use	Time Until Evaluation*	Quality of Extraction
Seinhorst elutriator	Density	500 mL‡	+++	+++	++	+++	15 min	++
Centrifugal flotation	Density	250 mL	++	+++	++	++	15 min	++
Wye washer	Density	1000 mL§	++	++	++	+++	15 min	++

+ = low, ++ = medium; +++ = high quality.

Besides, the method recommended for various groups of nematodes varies based on the type of study. For example, root-knot nematode can be extracted from roots and soil samples. On the other hand, rhabditid nematodes are extracted from soil, except for EPN, for which a Galleria bait trap is needed. The appropriate method for various groups of nematodes is summarized in Table **2**.

Table 2. Methods recommended for the extraction of nematodes from various sources [17, 33].

-	Direct Examination	Baermann Funnel/ Oostenbrink Dish	Root Incubation	Mistifier Technique	Maceration and Filtration	Maceration and Centrifugal Flotation	Galleria Bait Trap	Tray Method
Seeds								
Aphelenchoides besseyi	-	X	-	-	-	-	-	X
Ditylenchus destructor/D. dispaci	-	X	-	-	-	-	-	X
Foliage	-	-	-	-	-	-	-	-
Aphelenchoides besseyi	X	X	-	X	-	-	-	X
Root								
Nacobbus aberrans	X	-	X	-	-	X	-	X
Meloidogyne species	X	X	-	-	-	X	-	X
Radopholus similis	X	X	-	-	-	-	-	X
Hirschmanniella spp.	X	X	-	X	X	X	-	X
Tuber/bulb								
Meloidogyne species	-	-	-	-	-	X	-	X
Nacobbus aberrans	X	X	-	-	X	-	-	-
Plant tissue								

(Table 2) cont.....

-	Direct Examination	Baermann Funnel/ Oostenbrink Dish	Root Incubation	Mistifier Technique	Maceration and Filtration	Maceration and Centrifugal Flotation	Galleria Bait Trap	Tray Method
Ditylenchus destructor/D. dispaci	-	X	-	-	-	-	-	X
Radopholus similis	-	-	-	-	X	X	-	X
Wood and wood products								
Bursaphelenchus xylophilus	-	X	-	X	-	-	-	X
Vector beetle								
Bursaphelenchus xylophilus	X	X	-	-	-	-	-	-
EPN	-	-	-	-	-	-	X	-
Soil								
Rhabditid	-	X	-	-	-	X	-	X
Plant-parasitic	-	X	-	-	-	X	-	X
Larvae of EPN	-	-	-	-	-	-	-	X

CONCLUSION

Nematode extraction is a crucial process in nematology research. The choice of extraction method depends on the objective of the study. For example, in the case of entomopathogenic nematode (EPN) studies, the extraction method is different from that of soil nematodes. For EPNs, the nematodes are usually extracted from live insects, while for soil nematodes, the extraction is done from soil samples. Similarly, for slow-moving nematodes such as criconematids, a suitable extraction method must be chosen. These nematodes are usually found deep in the soil and require a gentle extraction process to avoid damage. The choice of extraction method is also influenced by the nematode's feeding group, which can be bacterial-feeding, fungal-feeding, plant-feeding, or omnivorous. Moreover, the extraction process also affects the quality of individual nematodes for molecular studies. In molecular biology research, fresh and live nematodes are necessary to obtain high-quality DNA or RNA samples. The extraction process should be gentle and quick to avoid any damage to the nematodes. Finally, it is essential to consider the feeding group of nematodes based on the study's requirements. Different feeding groups have different ecological roles and, therefore, differ in their distribution and abundance in soil. This information is essential for designing an appropriate sampling strategy to obtain representative samples for the study.

CHAPTER 4

Nematode Observations

Abstract: Preservation of nematodes is a crucial element of research endeavors, and it can be accomplished either temporarily or permanently. In order to conduct a rapid examination, researchers may place a minute droplet of water onto a glass slide for temporary observation. However, for the purpose of long-term preservation and collection, permanent slides are requisite. Prior to studying nematodes under a microscope for taxonomic purposes, it is necessary to first kill and fix them. This method is particularly useful for taxonomic studies involving light, electron (SEM), or transmission electron (TEM) microscopy.

Keywords: Fixative, Glass slide, Permanent, Preserving, Temporarily.

FORMALIN-GLYCERINE METHOD

Formalin harms humans; however, most fixatives include formalin to kill the nematode [10]. It should be noted that dehydration is critical for nematode preservation [10].

Materials

A volume of 100 ml can be prepared for each of the three fixative solutions:

Fixative I: 88 ml distilled water, 10 ml formaldehyde (40%), 1 ml acetic acid, 1 ml glycerine.

Fixative II: 95 ml ethanol (96%), 5 ml glycerine.

Fixative III: 50 ml ethanol (96%), 50 ml glycerine.

Procedure

1. Remove the additional water from the petri dish (in which the nematodes are to be suspended (Fig. **1A-H**) using soft paper (Fig. **1B**), and soak it until only a thin layer of water is present in the petri dish (Fig. **1C**).

2. Heat Fixative I to approximately 80 ˚C (Fig. **1E**).

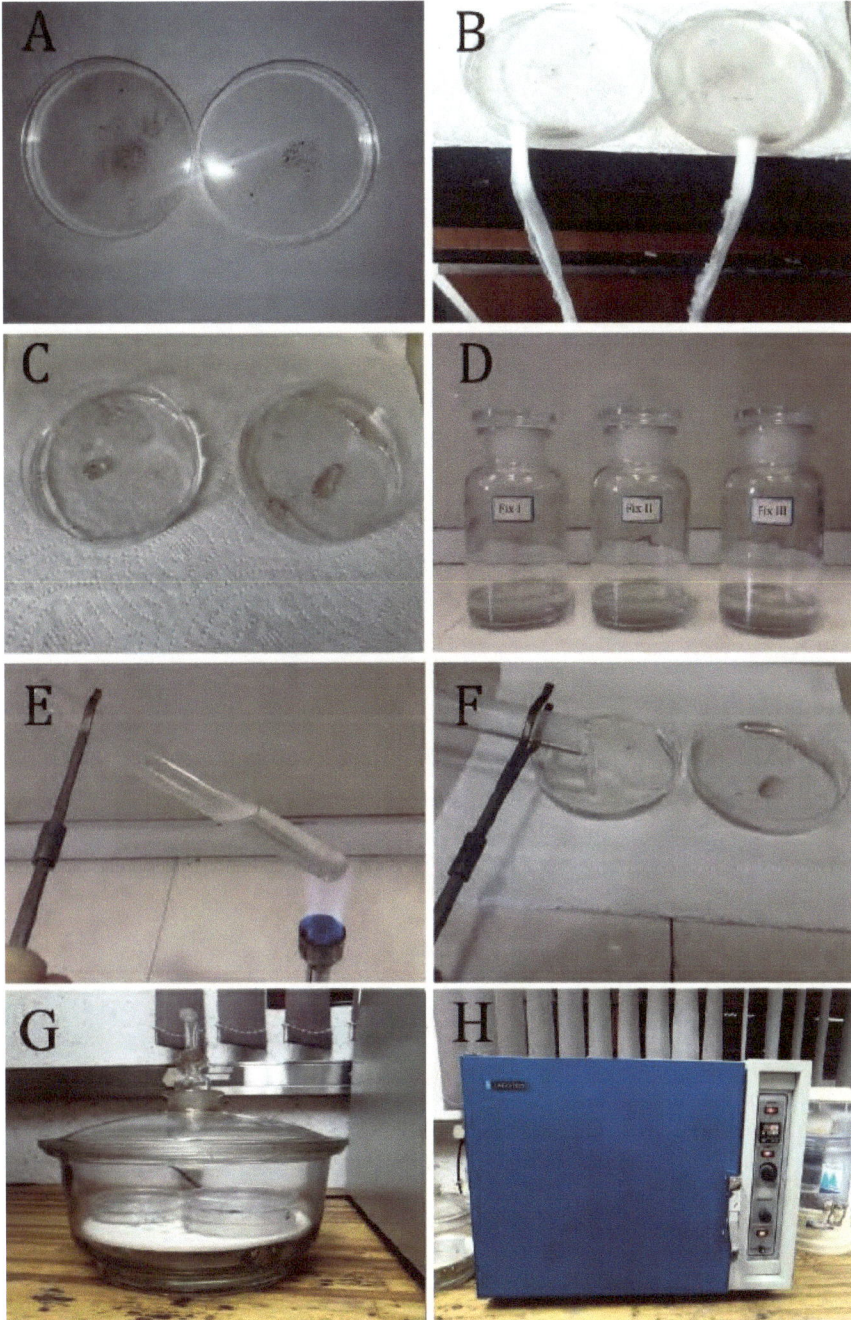

Fig. (1). Fixation of the nematodes. **A**: Nematodes in the water solution in the petri dish. **B**: Removing the water from the petri dish using soft paper. **C**: Petri dish with a thin layer of water after removing redundant water. **D**: Fixative I-III. **E**: Heating fixative I. **F**: Pouring the fixative into the petri dish. **G**: Desiccator with nematodes covered by glycerine after fixation processing. **H**: Incubator for fixation processing.

3. Add warm Fixative I to the petri dish in which the nematodes are suspended (Fig. **1F**).

4. Put the petri dish that contains the nematodes suspended in Fixative I into a desiccator, the bottom of which is filled with 96% ethanol (Fig. **1G**).

5. Put the desiccator into an incubator at 37 °C for 18-24 hr (Fig. **1H**).

After 18-24 Hours

6-Take out the petri dish from the desiccator.

7-Remove Fixative I by using soft paper to soak it up.

8-Add Fixative II to the petri dish.

9-Put the petri dish back into the desiccator.

10-Put the desiccator into the incubator for only 4 hrs (this step is critical).

After 4 Hours

11-Remove the desiccator from the incubator, take the petri dish with nematodes suspended in Fixative II out of the desiccator.

12-Add Fixative III to the petri dish or remove Fixative II, then add Fixative III.

13-Put the petri dish into the incubator (not into the desiccator) for at least 4 hrs until all the ethanol has evaporated.

Finally, all nematodes are suspended in glycerine to prepare permanent slides.

TAF (Triethanolamine Formalin) Method

This method is frequently used by many researchers and nematologists to preserve nematodes [2]. However, long-term storage of specimens in TAF leads to cuticle degradation [2, 34]. The procedure is as follows [2]:

1) Prepare TAF fixative (8% formalin and 2% triethanolamine in distilled water).

2) Transfer live nematodes to a small glass vial or Petri dishes and allow them to settle to the bottom. Draw off additional water until they are left in about 2 ml water.

3) Kill the nematodes by stirring the vial for 20-30 secs in a 70-90 °C water bath until they are dead and stretched.

4) Add an equal volume of 65-70 °C fixative. Stir, and then leave the vial alone for a day to allow the fixative to penetrate and act on all internal nematode organs.

5) Take the vial with TAF-fixed worms; remove the fixative under the stereomicroscope.

6) Fill the vial or block with a solution of 5% glycerin in distilled water. If the nematodes are in a vial, the solution should be over the nematodes.

7) Place the block or vial in an incubator at 35-40 °C and cover it nearly entirely. Then, leave them for slow evaporation until the water evaporates. Refill with 5% glycerin, then leave at least two more days in the incubator until all water evaporates. Then, the cuticle must be checked for its degree of dehydration. If the cuticle collapses, they are not yet wholly dehydrated; the process must be continued until the cuticle is completely firm for one or two more days. However, evaporation might take 4-6 days [35].

Slide Preparation

For classical studies, nematode specimens can be studied after being mounted on a temporary and permanent slide. Slide preparation is critical because only specimens in a good state can be used for identification.

Benefits of Temporary Slides

This method is used when the genus or species of a specific nematode needs to be determined quickly after extraction. On the other hand, by using temporary slides, the students and researchers can observe live specimens at a higher magnification after being observed under the stereomicroscope.

Procedure.

1. Add a drop of water to a glass microscope slide.

2. Transfer the nematode specimen to the center of the drop of water.

3. Cover the drop of water with a coverslip and seal the coverslip's edges with colorless nail polish. The slides can be used as soon as the nail polish has been hardened.

Specimens can also be mounted in lactophenol on temporary slides, which can be conserved longer than those mounted in water. They are used for mounting perineal patterns and sometimes whole nematodes. For the latter, specific structures may start fading after a few weeks. For this purpose, transfer the fixed nematodes to a drop of warm lactophenol on a glass slide and leave it on a hot plate for half an hour (30°C). Cover the specimen with a coverslip and seal the edges of the coverslip with nail polish.

Benefits of Permanent Slides

Permanent slides are crucial for taxonomical and even biodiversity studies. Using this method, slides can be used and stored for long periods. Glass slides and/or Cobb slides can be used. However, glass slides can break easily during handling, and the nematode specimen will be lost, while the Cobb slide with an aluminum frame is less prone to breakage.

Procedure

1. Wipe a glass slide with a soft cloth that has been dipped in 96% ethanol and put it onto the object table of a stereomicroscope (Fig. **2A-H**).

2. Take a glass or metal tube and heat the round edges of the open end.

3. Press the heated tip of the tube into paraffin wax.

4. Press the paraffin-wax-covered edge of the tube onto the glass slide prepared in step 1 to form a circle.

5. Clean the inner part of the paraffin-wax circle on the glass slide with an earbud/tip of tissue dipped in 96% ethanol to avoid the presence of wax droplets.

6. Transfer a small drop of pure glycerine to the middle of the wax circle on the glass slide using the fine tip of a needle.

7. Transfer the nematode specimen to be studied to the bottom of the drop of glycerine using a needle.

8. Place a coverslip onto the paraffin wax ring.

10. Heat the glass slide, using a hot plate or alcohol burner, to melt the paraffin gradually.

11. When the paraffin wax has solidified, seal the edges of the coverslip using transparent nail polish.

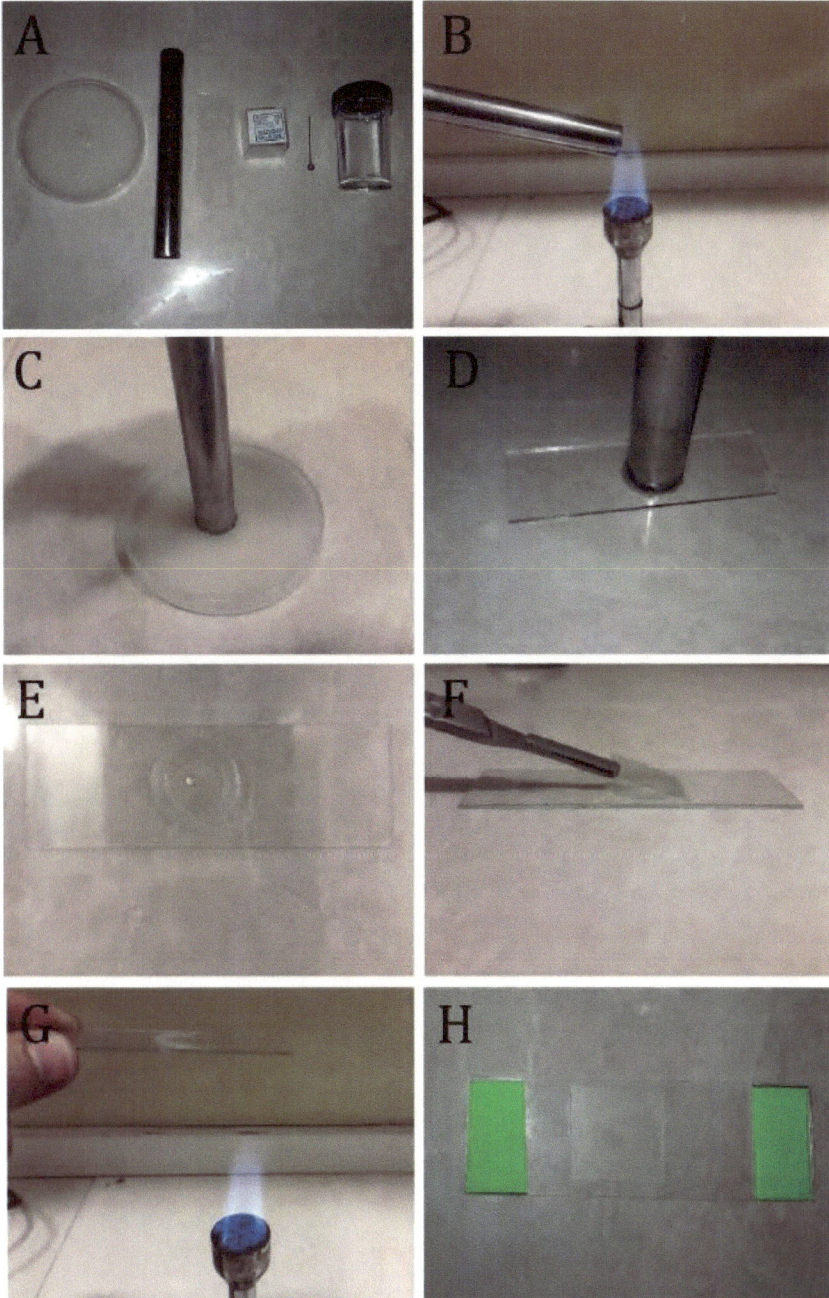

Fig. (2). Slide preparation. **A**: Materials for slide preparation. **B**: Warming up the cylinder. **C**: Pressing the cylinder into the petri dish containing paraffin wax. **D**: Making the ring of paraffin wax. **E**: Paraffin wax ring with a drop of glycerine in the middle. **F**: Covering the paraffin ring with a coverslip. **G**: Warming up the paraffin ring covered by the coverslip. **H**: Permanent slide.

CONCLUSION

The preservation of nematodes in glycerine is of utmost importance when it comes to studying their morphology. With proper preservation, researchers can obtain a clear and accurate view of the internal structures of various groups of nematodes, which is essential for taxonomical studies. This process enables researchers to identify and classify different species of nematodes accurately, furthering our understanding of these tiny yet significant organisms.

CHAPTER 5

Nematode Morphological Observations

Abstract: In the field of nematology, a variety of microscopes are utilized for distinct purposes, including standard light microscopes and stereomicroscopes. Additionally, line illustrations are essential for the taxonomical examination of nematodes. As such, it is crucial to properly calibrate and maintain microscopes to ensure accurate outcomes. This section delves into the essential duties associated with microscope usage in nematological research. Differential interference contrast microscopy (DIC) is a useful technique to create a high-quality photograph of the nematodes. In regard to taxonomy and diagnosis of nematodes, accurate measurements are of utmost importance. Each nematode family, whether they are free-living or plant-parasitic, necessitates specific measurements of particular attributes. The identification of nematodes relies on the de Man indices. This section offers visual aids for different families to showcase the significant features that must be measured. The art of capturing images of nematodes holds immense significance in the fields of taxonomy and histological research. Nematology employs numerous methods to obtain top-notch visuals for analysis, among which is Scanning Electron Microscopy (SEM). Furthermore, phase contrast is widely used in nematology for a variety of purposes. This chapter endeavors to provide a comprehensive understanding of the microscope, measurements, DIC, and SEM techniques.

Keywords: Calibration, DIC, Microscope, SEM.

MICROSCOPE JUSTIFICATION

Microscope Types

After slide preparation, the nematologist always uses a microscope to study nematode specimens for identification purposes since these organisms are microscopic. Since all nematologists use a microscope, the quality of the microscope is essential. Two types of microscopes are used for research work: a dissecting/stereomicroscope and a light microscope. The "stereomicroscope" (Fig. **1**) has a limited number of lower magnification rates (10-80x), offering a wide-angled vision. This microscope is mainly used for analyzing nematode samples while counting at only the genus/family level and/or for dissection work.

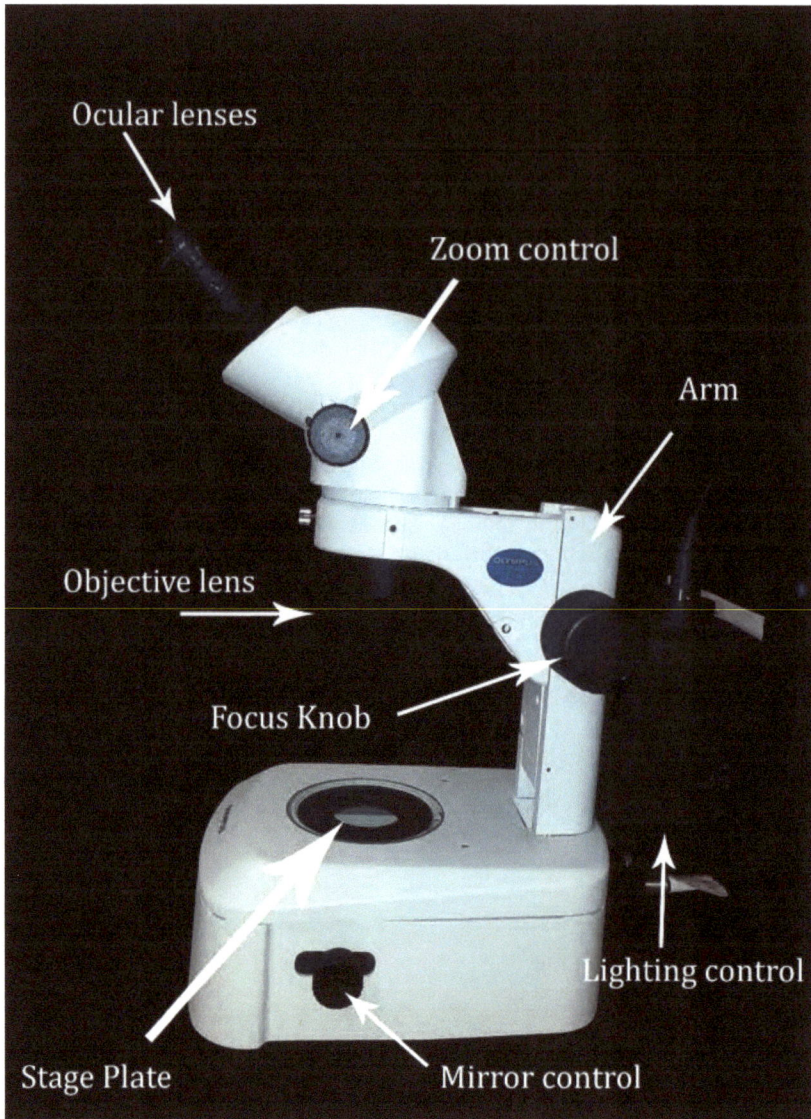

Fig. (1). Stereomicroscope and its different parts.

The light microscope (Fig. **2**) offers magnifications ranging from 10 to 1000x. Working with it at a magnification of 500 or 1000x requires an oil immersion lens. This type of microscope is generally used for the identification of nematodes. Usually, the nematodes studied with a light microscope are mounted using a dissecting microscope.

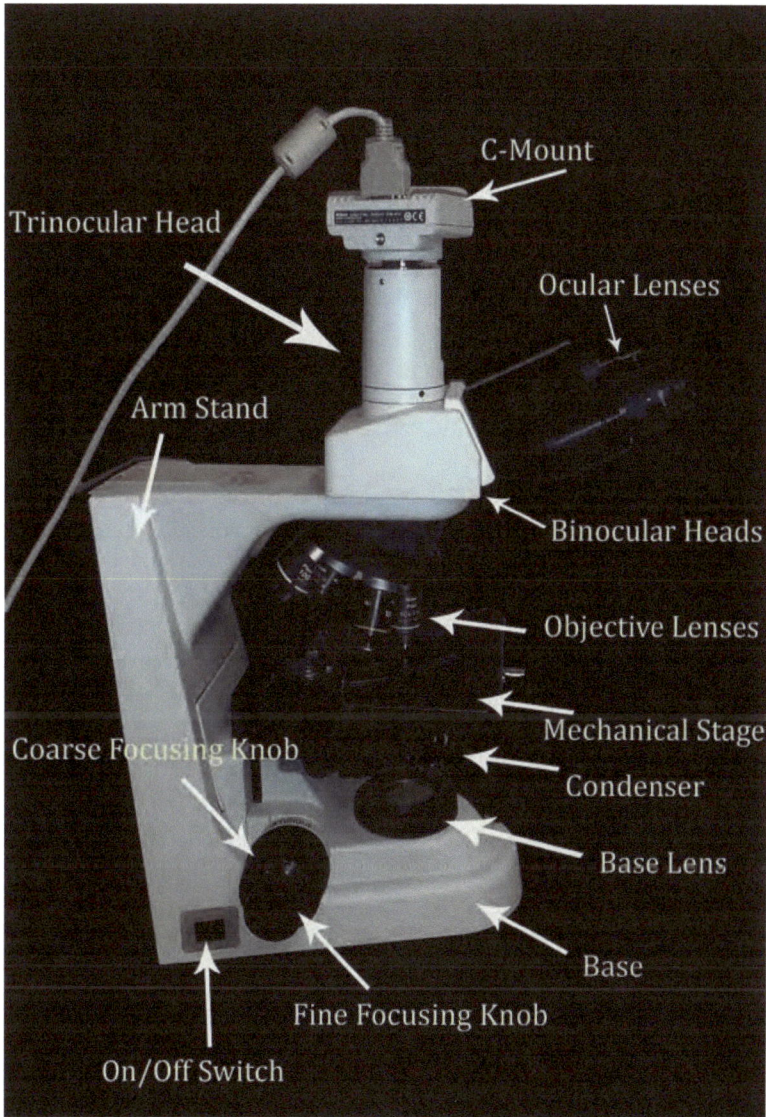

Fig. (2). The microscope and its different parts.

Calibrating the Eyepiece Graticule

To ensure accurate measurements of nematode specimens, the eyepiece graticule (Fig. **3**) on a microscope must be calibrated prior to use. The following steps should be taken in order to properly calibrate the eyepiece graticule:

EYEPIECE MICROMETER

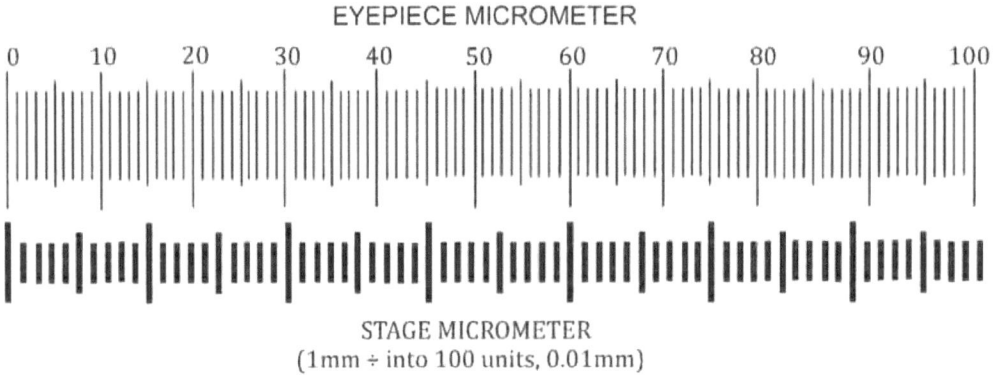

STAGE MICROMETER
(1mm ÷ into 100 units, 0.01mm)

Fig. (3). Eyepiece micrometer and its calibration.

Firstly, if the microscope does not already have a graticule, one must be inserted into the eyepiece by unscrewing the top lens, placing the graticule on the rim halfway down, and then replacing the top lens.

Next, a stage micrometer slide should be placed on the microscope's stage. It should be noted that the smallest division on the stage micrometer is equivalent to 100 μm. Using the low-power objective, the microscope should be focused on the stage micrometer. The eyepiece should be rotated, and the slide should be moved to align the scales of the eyepiece graticule and the stage micrometer. The number of divisions on the eyepiece graticule equivalent to 100 μm on the stage micrometer should be counted. From this, the length that one eyepiece division is equivalent to can be calculated. For instance, if three divisions are equal to 100 μm, then each division is equal to 33.3 μm at low magnification. Observations should be recorded. This process should be repeated for the medium. and high-magnification objectives. The eyepiece graticule should be calibrated before being used to measure nematode specimens' characteristics, ensuring accurate and reliable results.

Calibrating the Microscope for Drawing

1. Equip the microscope with a drawing tube (Fig. **4**) [10].

2. To begin, position the stage micrometer slide onto the stage of the microscope, ensuring it is properly aligned. Take a moment to observe the smallest division on the micrometer, which is equivalent to 100 micrometers (μm). This information is crucial for making accurate measurements during your observations.

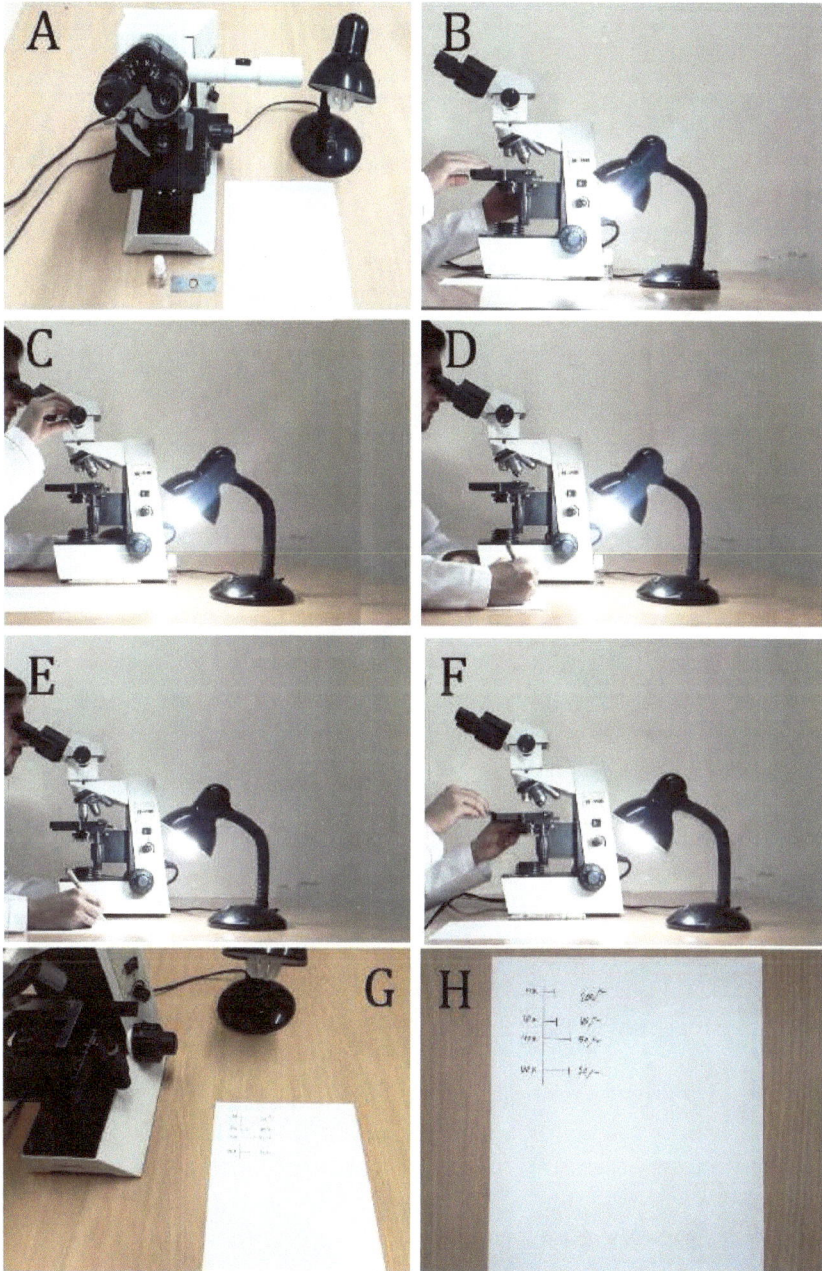

Fig. (4). Calibration of a microscope equipped with a drawing tube. A: Microscope plus a drawing tube. B: Inserting the calibrating slide into the stage. C: Adapting the ocular lens with the calibrating slide. D-G: Adapting the different magnifications with the calibrating slide using lines on the paper. H: Calibrated magnifications using lines and relative micrometers.

3. To accurately measure the specimen under the microscope, we need to use the low-power objective lens and focus it on the stage micrometer. Next, draw a line on the object with a pen and adjust it to precisely 100 µm. This will help us obtain accurate and reliable results from our observation.

4. To calibrate the microscope for all magnifications, repeat step 3 for medium and high magnification objectives. This will allow you to measure any characteristic of the nematode specimen.

It is better to adjust the line's length on the paper from 100 to 50 µm for 4x to 50x magnifications, respectively. For 100x, the line's length can be adjusted for 10, 25, or 50 µm.

However, nowadays, the drawings are based on the LM photos taken by a digital camera and processed by Photoshop or CorelDraw software. The old method of drawing is more laborious, but eventually, it can be more exciting if the drawing is done with passion.

Maintenance of microscopes

When transporting microscopes, they should be packed carefully in their original containers/proper containers with surrounding protective packaging materials to prevent them from moving. Put away the microscope from anything that can wet the lens and electric parts and place it away from direct sunlight. Always handle the microscope and its parts carefully to ensure optimal and efficient functioning thereof. View the specimen with 4x, 10x, and 40x, and then with the 100x. The specimens under 100x should always be handled with an immersion oil. To clean the lens, use a proper cloth or tissue designed for the microscopes.

Differential Interference Contrast (DIC)

Differential interference contrast (DIC) is a unique microscope method that contrasts the images with little or no focus in brightfield microscopy. The DIC microscope provides a 3-D photograph, which is critical for preparing high-quality pictures for taxonomy purposes. Additionally, DIC is ideal for the unstained nematode species. DIC microscopes use an infrared (IR) light source, which is suitable for imaging thick specimens of nematode species. This is because of the longer wavelength for light entering the tick tissues. Hence, the DIC microscope is better than phase or oblique contrast for taxonomical studies. A reason for that is a halo artifact that does not exist in the DIC microscope. The picture below (Fig. **5**) shows the process of the light for imaging through DIC [36]. In addition, DIC provides a better photograph for nematode taxonomy compared to other microscope types (Fig. **6**).

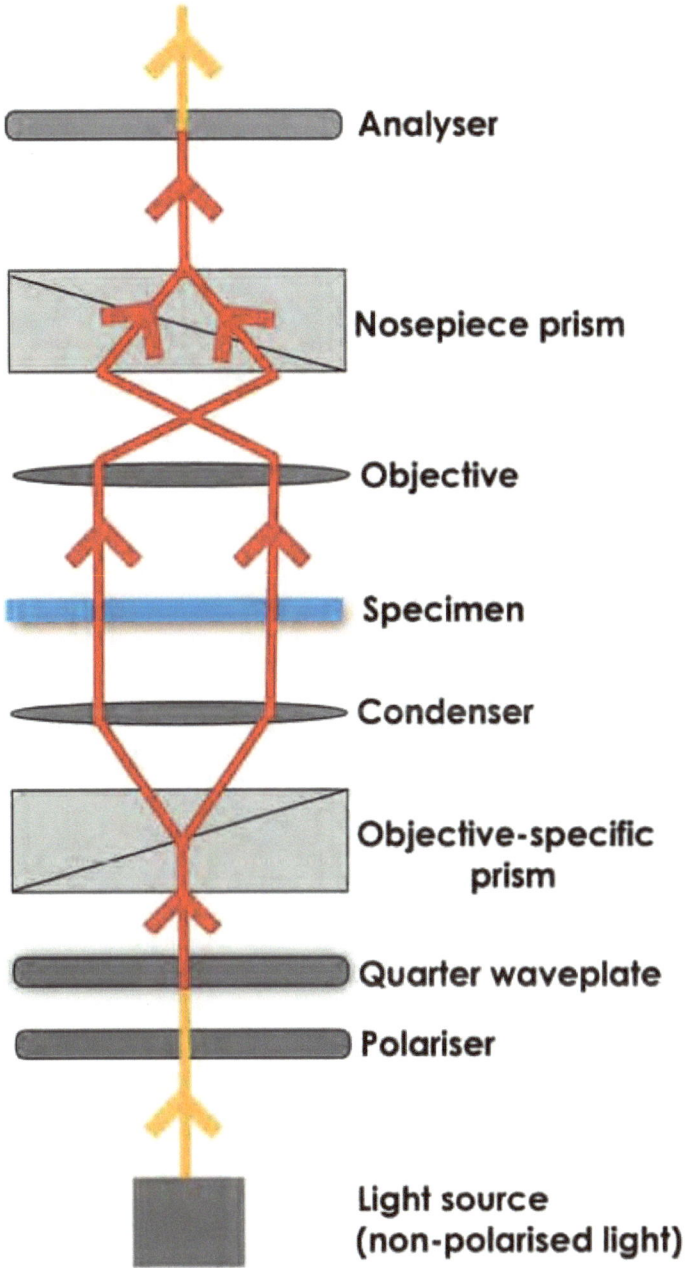

Analyser

Nosepiece prism

Objective

Specimen

Condenser

Objective-specific
prism

Quarter waveplate

Polariser

Light source
(non-polarised light)

Fig. (5). A diagram showing the light processing in DIC microscopes (adapted from Scientifica [36]).

NEMATODE MEASUREMENTS

Various measurements of nematodes' bodies and associated structures are needed for describing nematodes [10]. Depending on the nematodes to be studied, the measurements can be different. However, the ratios found by De Man [37] are constant for all nematodes and useful for taxonomical purposes. These ratios are listed below:

Ratios for all groups of nematodes [31] (the indices should be explained in the illustration).

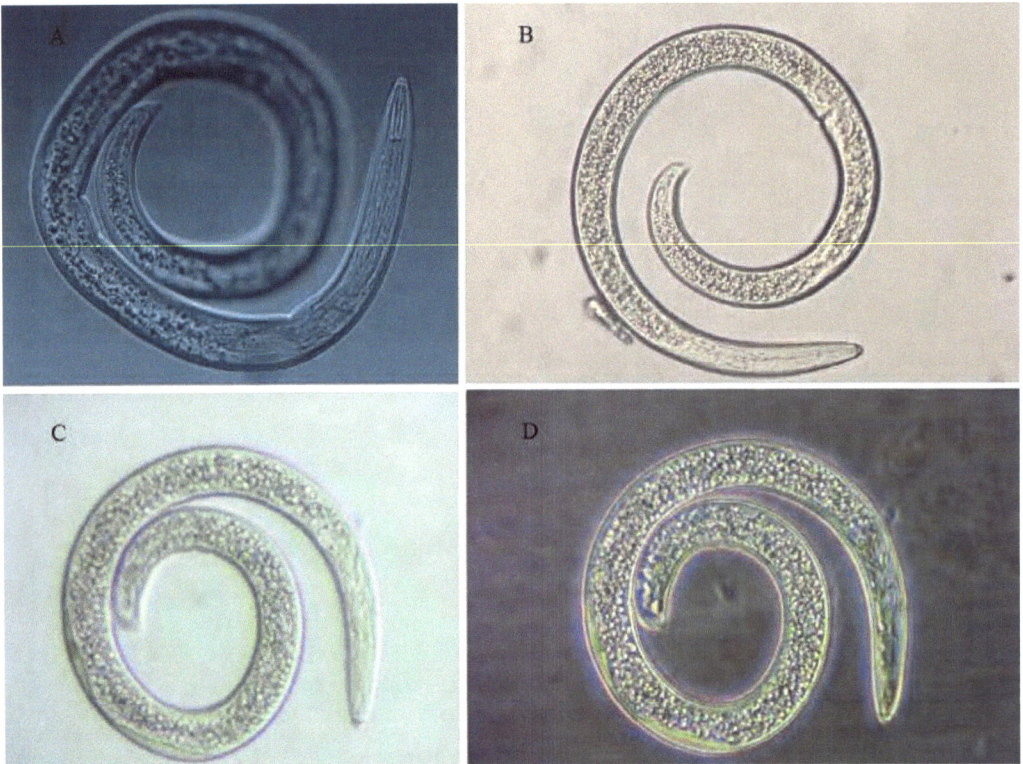

Fig. (6). A comparison of DIC, Bright field, and Phase contrast photo of *Helicotylenchus*. A: DIC. B, C: Bright field. D: Phase contrast. (Photo by: Ebrahim Shokoohi).

L = Total body length in microns or millimeters.

a = Body length ÷ greatest width (generally at vulva region).

b = Body length ÷ pharynx length.

b' = Body length ÷ pharynx length from the lips to the end of the pharyngeal gland.

c = Body length ÷ tail length.

c' = Tail length ÷ body width at the anus.

V = Distance of the vulva from the lips ÷ body length × 100.

G1 = Overall length of the anterior ovary from vulva ÷ body length × 100.

G2 = Overall length of the posterior ovary from vulva ÷ body length × 100.

Ratios mainly for plant-parasitic nematodes of Tylenchomorpha: (Siddiqi, 2000).

T= distance from cloacal aperture to the anterior end of testis ÷ body length × 100.

m = length of conus as a percentage of total stylet length.

O = distance between stylet base and orifice of the dorsal pharyngeal gland as a percentage of stylet length.

MB = distance between the anterior end of the body and the center of the median pharyngeal bulb as a percentage of the pharyngeal length.

R = total number of body annulus.

Roes = number of the annulus in the pharyngeal region.

Rex = number of the annuals between the anterior end of the body and excretory pore.

RV = number of the annuli between the posterior end of the body and the vulva.

RVan = number of the annuals between the vulva and anus.

Ran = number of the annulus on the tail.

VL/VB = distance between the vulva and posterior end of the body (tail tip) ÷ by body width at the vulva.

In dorylaim nematodes, Loof and Coomans [38] added several sets of morphometric characters in which there are five pharyngeal glands in relation to the length of the pharynx as a percentage. The glands and their orifice are among the characters [39], including:

DO = orifice of the dorsal gland.

DN = dorsal gland nucleus at the centre of the nucleolus.

S_1O = orifice of the first pair of ventrosublateral glands.

S_1N_1 = anterior nucleus of the first pair of ventrosublateral glands.

S_1N_2 = posterior nucleus of the first pair of ventrosublateral glands.

S_2O = orifice of the second pair of ventrosublateral glands.

S_2N_1 = anterior nucleus of the second pair of ventrosublateral glands.

S_2N_2 = posterior nucleus of the second pair of ventrosublateral glands.

K = Distance $DN\text{-}S_1N_1$ as a percentage of distance $DN\text{-}S_1N_2$.

K' = Distance $DO\text{-}S_1O$ as a percentage of DOS_2O.

Pictorial Measurement Guide

Characters that should be measured vary among the different groups of nematodes. For instance, plant-parasitic nematodes of the infraorder Tylenchomorpha bear a stylet, while free-living Rhabditida has a stoma. Measurements of various characteristics of a nematode specimen are demonstrated with colored lines in Figs. (**7 - 15**).

SCANNING ELECTRON MICROSCOPY (SEM)

Scanning electron microscopy is a specialized type of microscope that can take a picture of an organism's surface. Basically, an SEM is composed of two parts: the electronic console and the electronic column. Nowadays, different types of SEMs are provided by other companies, *e.g.*, Phenom, which takes up less space and is very easy to operate. To take a photograph with an SEM, the nematode specimens should be prepared using a specific method. SEM photographs are beneficial for the accurate identification of nematodes, especially for free-living nematodes (family Cephalobidae), which have 'labial probolae' and plant-parasitic nematodes. With SEM pictures, characters on the surface of the nematode body are visible. For example, in some species, the lateral field is obscure when viewing it with a light microscope, while with an SEM, the real shape and incisures line is visible. This characteristic is undeniable in some genera, viz. *Ditylenchus*. Abolafia [40] suggested a basic and cheap container for preparing nematode specimens for SEM studies. The protocol for SEM preparation is summarized below.

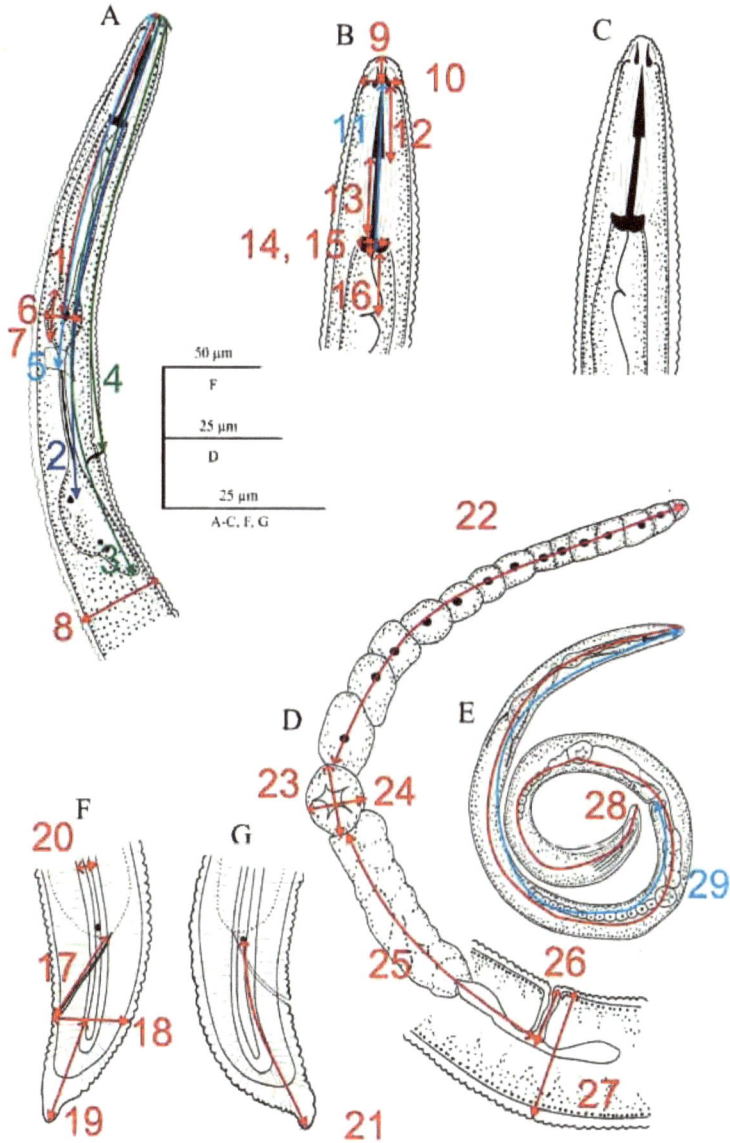

Fig. (7). The demonstration, with colored lines, of the way various characteristics of nematode specimens belonging to Tylenchoidea. 1. Anterior end to the median bulb. 2. Anterior end to the dorsal gland. 3. Neck length. 4. Anterior end to Excretory pore. 5. Anterior end to nerve ring. 6. Median bulb length. 7. Median bulb width. 8. Body diameter at the end of the neck. 9. Lip region length. 10. Lip region width. 11. Stylet length. 12. Conus. 13. Shaft. 14. Knobs length. 15. Knobs width. 16. DGO. 17. Rectum. 18. Anal body diameter. 19. Tail length. 20. Lateral field width. 21. Phasmid to the posterior end. 22. Ovary length. 23. Spermatheca length. 24. Spermatheca width. 25. Uterus length. 26. Vagina length. 27. Body diameter at the vulva. 28. Body length. 29. Anterior end to the vulva.

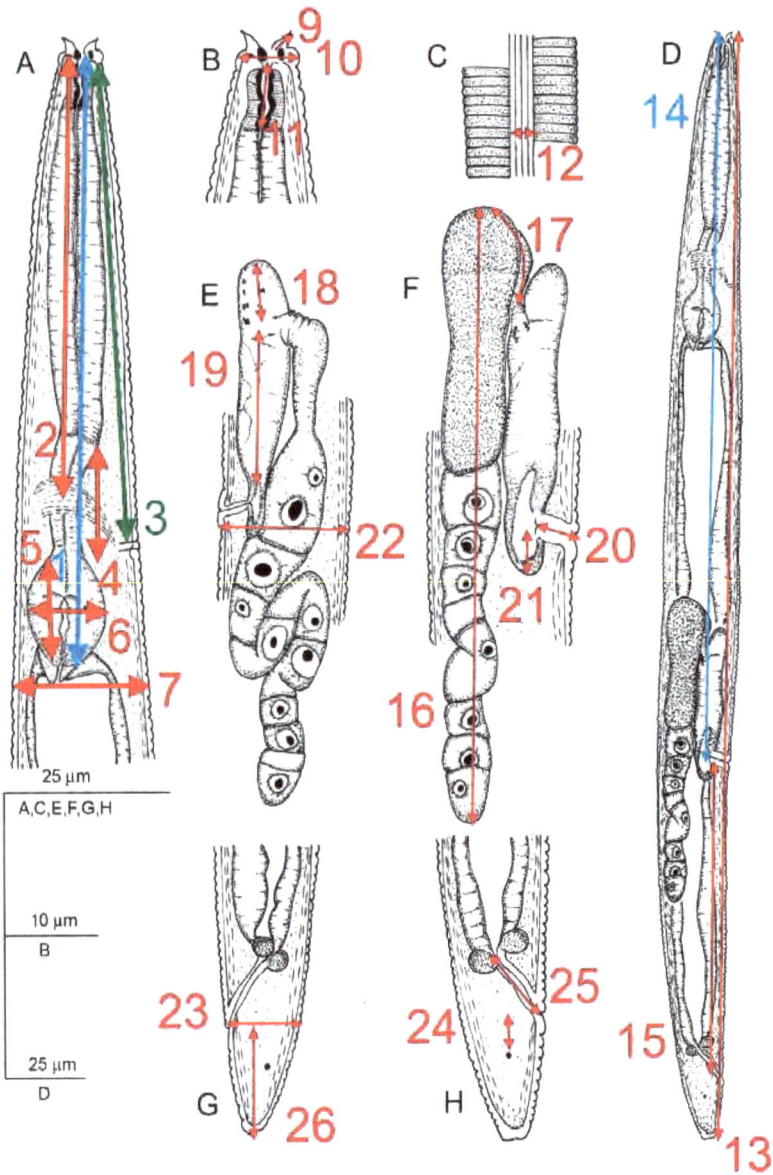

Fig. (8). The demonstration, with colored lines, of the way various characteristics of nematode specimens belonging to Cephaloboidea. 1. Neck length. 2-Anterior end to nerve ring 3. Anterior end to Excretory pore 4. Isthmus length. 5. Basal bulb length. 6. Basal bulb width. 7. Body diameter at the end of the neck. 9. Probolae length. 10. Lip region width. 11. Stoma width. 12. Lateral field width. 13. Body length. 14. Anterior end to the vulva. 15. Vulva to anus distance. 16. Ovary length. 17. Oviduct. 18. Spermatheca length. 19. Uterus length. 20. Vagina length. 21. Post vulval uterine sac length. 22. Body diameter at the vulva. 23. Anal body diameter. 24. Phasmid to anus distance. 25. Rectum. 26. Tail length.

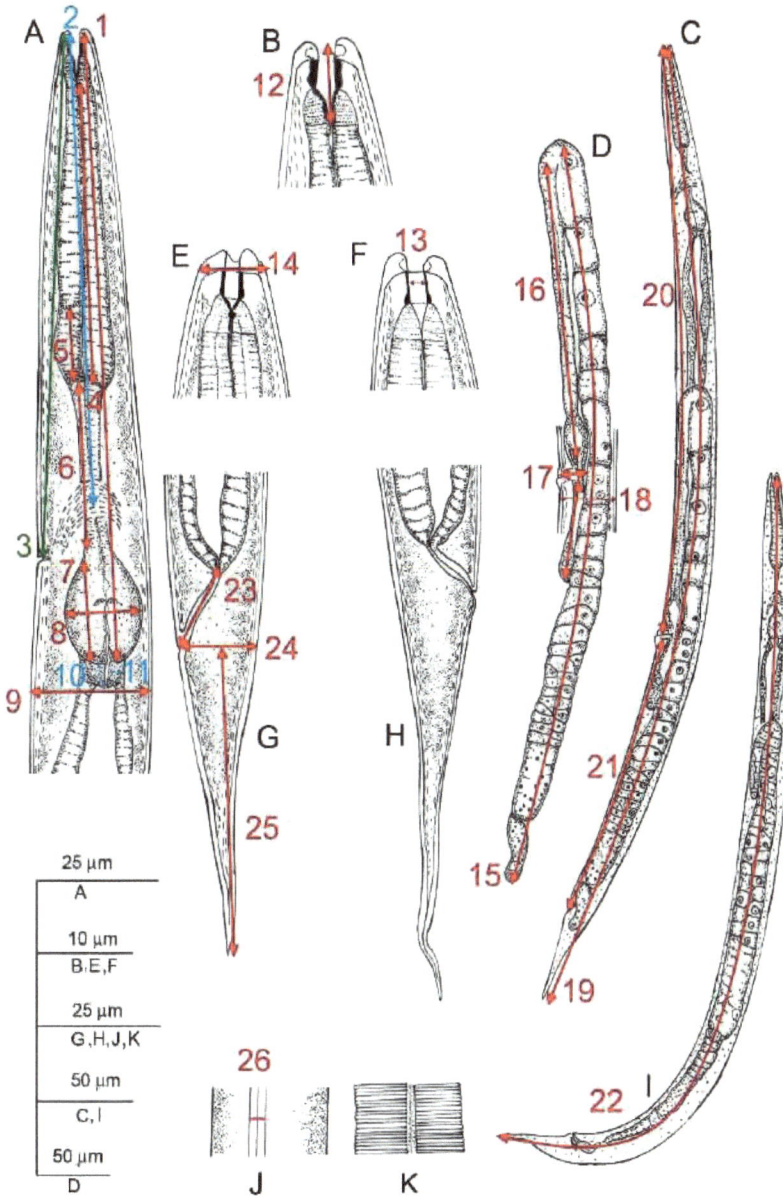

Fig. (9). The demonstration, with colored lines, of the way various characteristics of nematode specimens belonging to Panagrolaimoidea. 1. Neck length. 2-Anterior end to nerve ring 3. Anterior end to Excretory pore 4. Procorpus length. 5-Metacorpus length. 6. Isthmus length. 7. Basal bulb length. 8. Basal bulb width. 9. Body diameter at the end of the neck. 10. Cardia length. 11. Cardia width. 12. Stoma length. 13. Stoma width. 14. Lip region width. 15. Ovary length. 16. Uterus length. 17. Vagina length. 18. Body diameter at the vulva. 19. Female length. 20. Anterior end to the vulva. 21. Vulva to anus distance. 22. Male body length. 23. Rectum. 24. Anal body diameter. 25. Tail length. 26. Lateral field width.

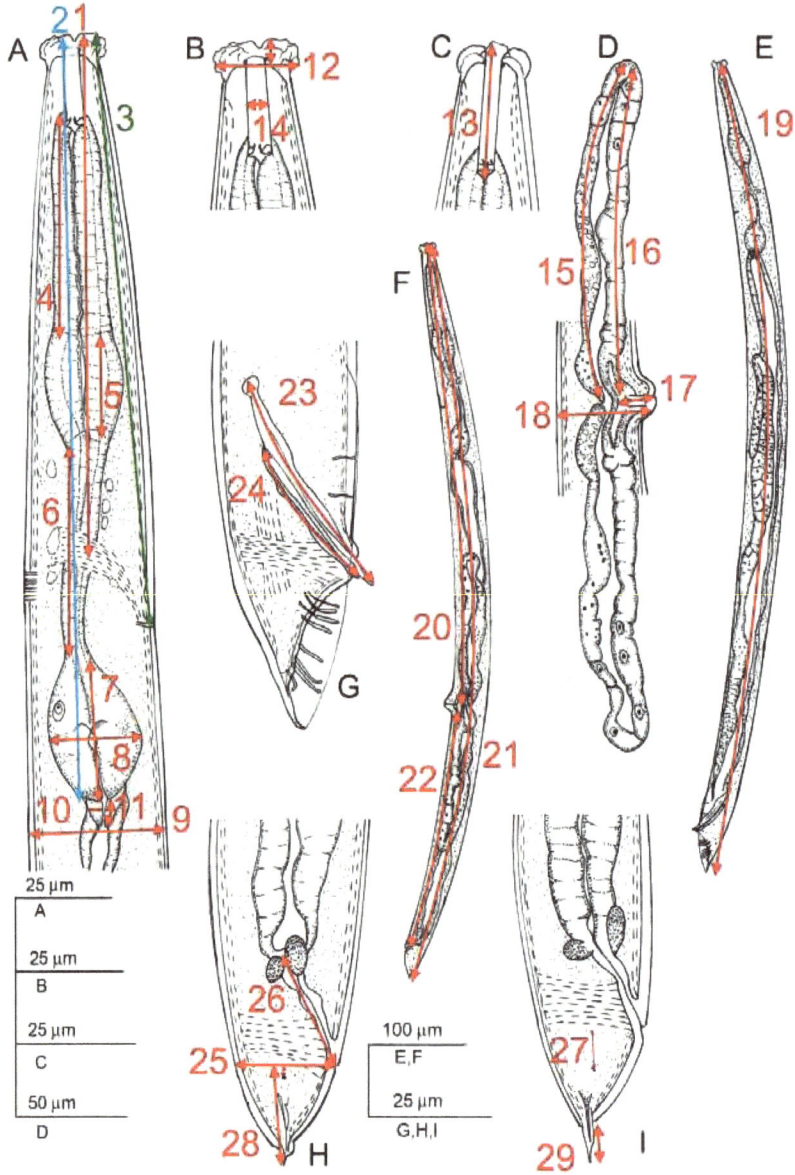

Fig. (10). The demonstration, with colored lines, of the way various characteristics of nematode specimens belonging to Rhabditoidea. 1. Anterior end to nerve ring. 2. Neck length. 3. Anterior end to Excretory pore 4. Corpus length. 5. Metacorpus length. 6. Isthmus length. 7. Basal bulb length. 8. Basal bulb width. 9. Body diameter at the end of the neck. 10. Cardia width. 11. Cardia length. 12. Lip region width. 13. Stoma length. 14. Stoma width. 15. Ovary length. 16. Uterus length. 17. Vagina length. 18. Body diameter at the vulva. 19. Male body length. 20. Anterior end to the vulva. 21. Female length. 22. Vulva to anus distance. 23. Spicule length. 24. Gubernaculum length. 25. Anal body diameter. 26. Rectum. 27. Phasmid to anus. 28. Tail length. 29. Tail projection.

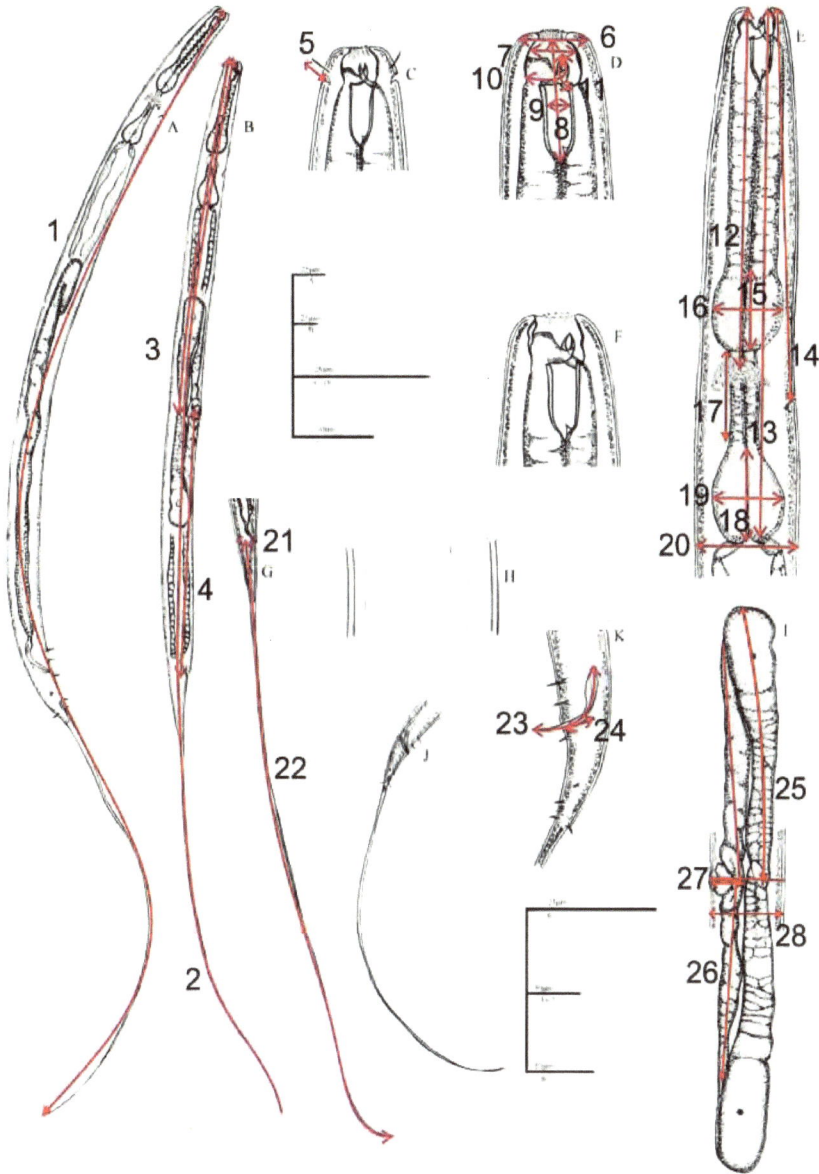

Fig. (11). The demonstration, with colored lines, of the way various characteristics of nematode specimens belonging to Diplogastroidea. 1. Male length. 2. Female length. 3. Anterior end to the vulva. 4. Vulva to the posterior end. 5. Setae length. 6. Lip region width. 7. (Gymnostom width). 8. Stoma length. 9. (stegostom width) 10. Large Tooth length. 11. Small tooth length. 12. Anterior end to nerve ring. 13. Neck length. 14. Anterior end to Excretory pore. 15. Metacorpus length. 16. Metacorpus width. 17. Isthmus length. 18. Basal bulb length. 19. Basal bulb width. 20. Body diameter at the end of the neck. 21. Anal body diameter. 22. Tail length. 23. Spicule length. 24. Gubernaculum length. 25. Ovary length. 26. Uterus length. 27. Vagina length. 28. Mid body diameter.

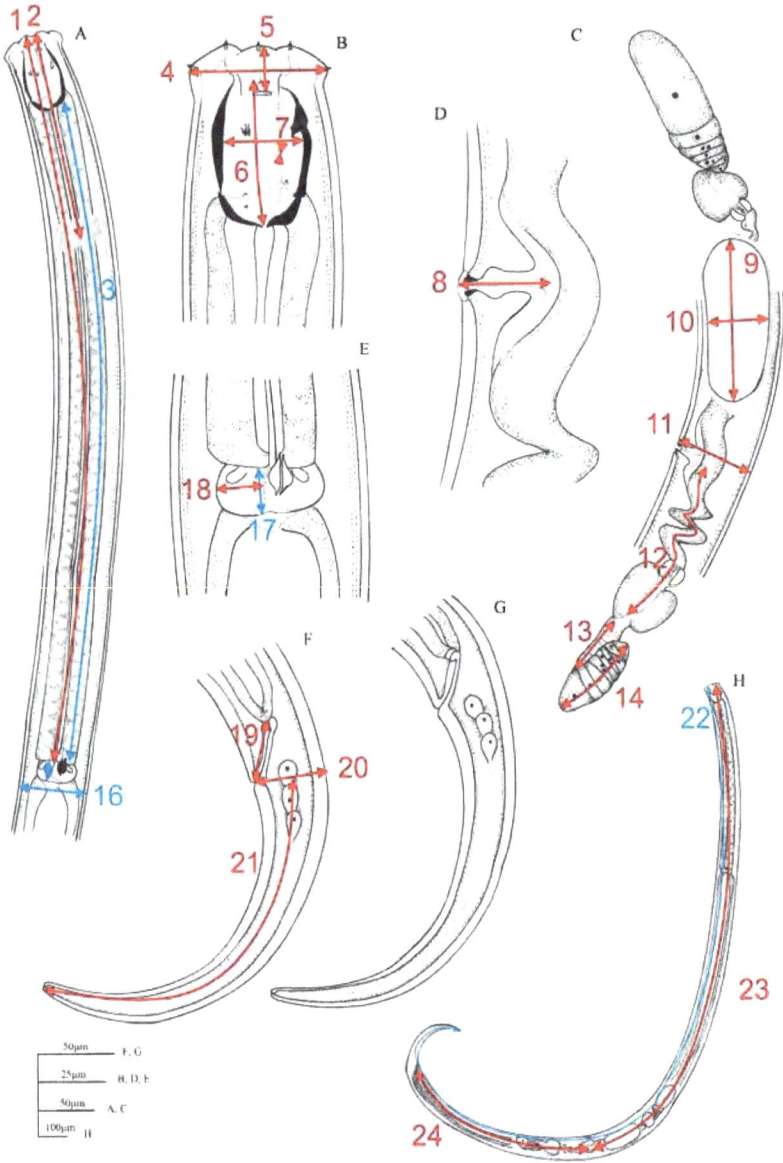

Fig. (12). The demonstration, with colored lines, of the way various characteristics of nematode specimens belonging to Mononchida. 1. Neck length. 2. Anterior end to nerve ring. 3. Pharynx length. 4. Lip region width. 5. Anterior end to Amphid. 6. Buccal cavity length. 7. Buccal cavity width. 8. Vagina length. 9. Egg length. 10. Egg width. 11. Body diameter at the vulva. 12. Uterus length. 13. Oviduct length. 14. Ovary length. 16. Body diameter at the end of the neck. 17. Cardia width. 18. Cardia length. 19. Rectum. 20. Anal body diameter. 21. Tail length. 22. Female length. 23. Anterior end to vulva. 24. Vulva to anus distance.

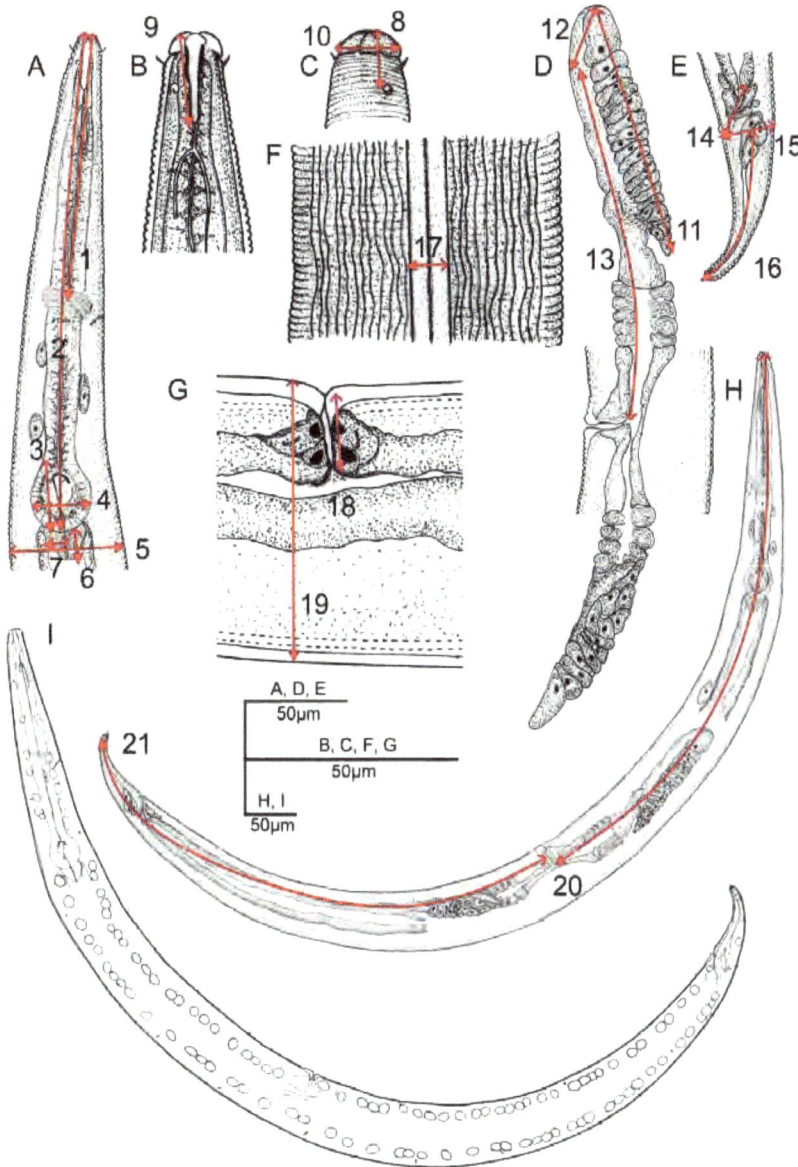

Fig. (13). The demonstration, with colored lines, of the way various characteristics of nematode specimens belonging to Plectida. 1. Anterior end to nerve ring. 2. Neck length. 3. Basal bulb length. 4. Basal bulb width. 5. Body diameter at the end of the neck. 6. Cardia length. 7. Cardia width. 8. Anterior end to Amphid. 9. Stoma length. 10. Lip region width. 11. Ovary length. 12. Oviduct length. 13. Uterus length. 14. Rectum. 15. Anal body diameter. 16. Tail length. 17. Lateral field width. 18. Vagina length. 19. Body diameter at the vulva. 20. Anterior end to the vulva. 21. Vulva to the posterior end.

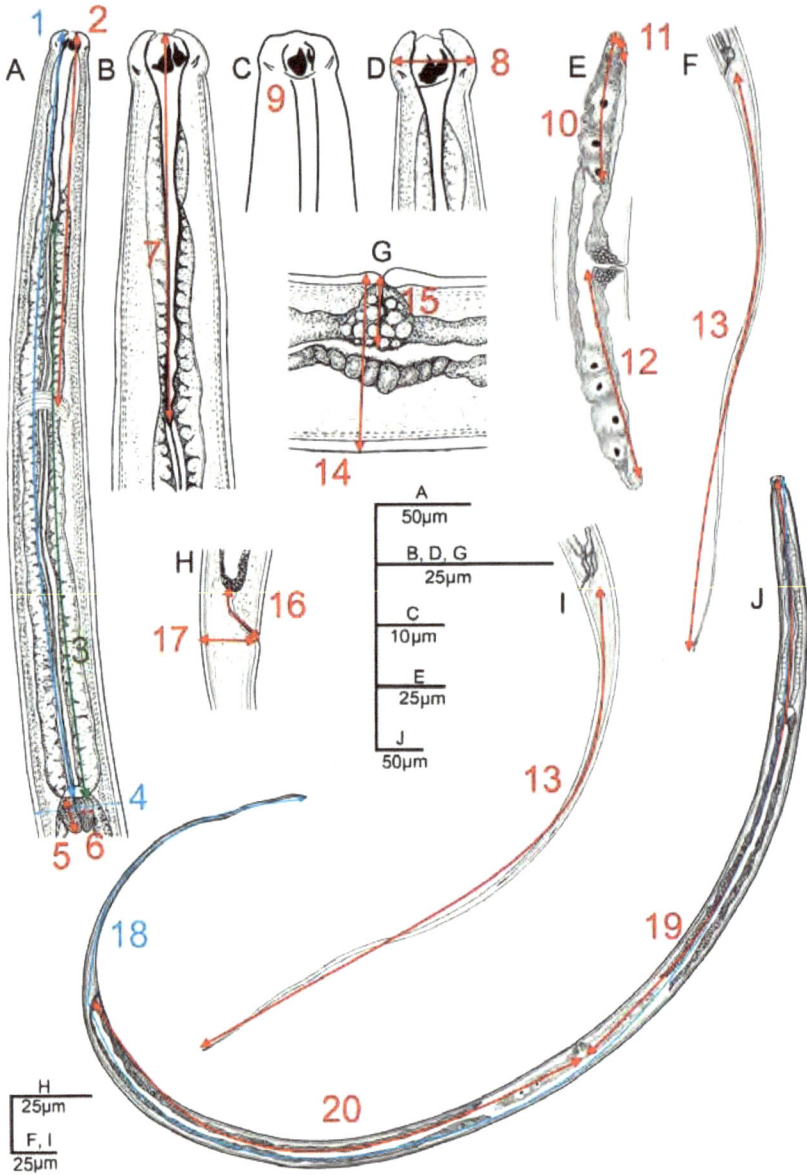

Fig. (14). The demonstration, with colored lines, of the way various characteristics of nematode specimens belonging to Enoplida. 1. Anterior end to nerve ring. 2. Neck length. 3. Pharynx length. 4. Body diameter at neck. 5. Cardia length. 5. Cardia width. 7. Stoma length. 8. Lip region width. 9. Tooth length. 10. Ovary length. 11. Oviduct. 12. Uterus length. 13. Tail length. 14. Body diameter at the vulva. 15. Vagina length. 16. Rectum. 17. Anal body diameter. 18. Female length. 19. Anterior end to the vulva. 20. Vulva to anus distance.

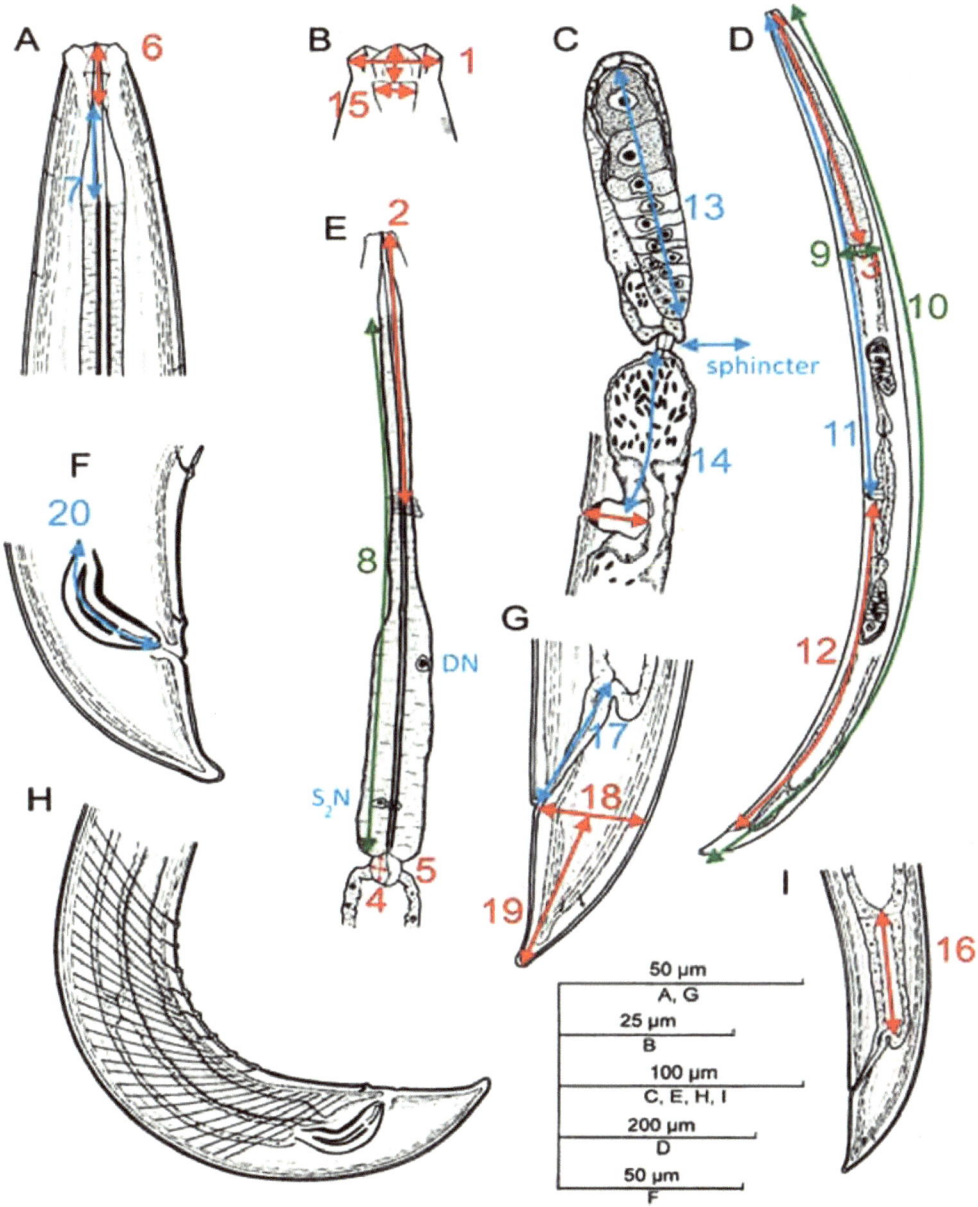

Fig. (15). The demonstration, with colored lines, of the way various characteristics of nematode specimens belonging to Dorylaimida. 1. Lip region width. 2. Anterior end to nerve ring. 3. Neck length. 4. Cardia length. 5. Cardia width. 6. Odontostyle length.7. Odontophore length. 8. Pharynx length. 9. Body diameter at neck base. 10. Female length. 11. Anterior end to the vulva. 12. Vulva to anus distance. 13. Ovary length. 14. Uterus length. 15. Amphid aperture width. 16. Prerectum length. 17. Rectum. 18. Anal body diameter. 19. Tail length. 20. Spicule length.

Preparation of Specimens for SEM Study

1-Remove the coverslip from the microscope slide on which a nematode was mounted, which was studied using a light microscope, and transfer the nematode to a drop of glycerin and gradually add a drop of water (Fig. **16**).

2-After two hours, add more water gradually and leave it to stand for one hour.

3-Transfer the nematode to distilled water and leave it standing for 30 min.

4-Transfer the nematodes to a tube that contains a 5% ethanol solution and leave it standing for one day.

5-Transfer the nematodes to a series of ethanol 30, 50, 70, 90, 95, and 100%. The nematodes should be kept in each concentration of ethanol for two hours.

6-Transfer the nematodes to a tube that contains pure acetone.

7-Put the nematodes in carbon dioxide (critical point drying).

8-Gold coat the nematodes in 12 mA for 15 min.

9-Study the samples coated with gold at 4 kV with an SEM (*e.g.*, JEOL JSM-5800).

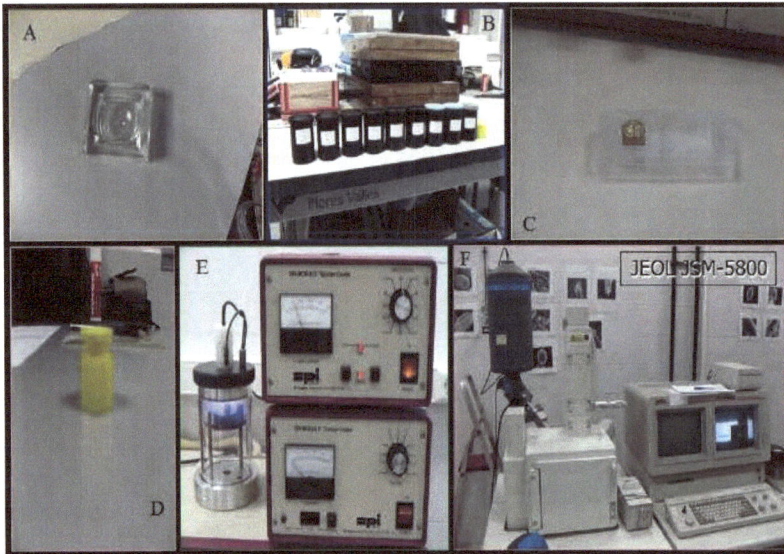

Fig. (16). Processing steps for SEM. A: Glass block with specimens. B: Ethanol series for nematode dehydration. C: Nematode samples coated with gold. D: Container for the samples. E: Instrument for gold coating of nematodes. F: Scanning Electron Microscope. (The figure is according to Shokoohi [10]).

A scanning electron microscope (SEM) image and a photograph taken by scanning microscopy are presented in Figs. (17 and 18), respectively.

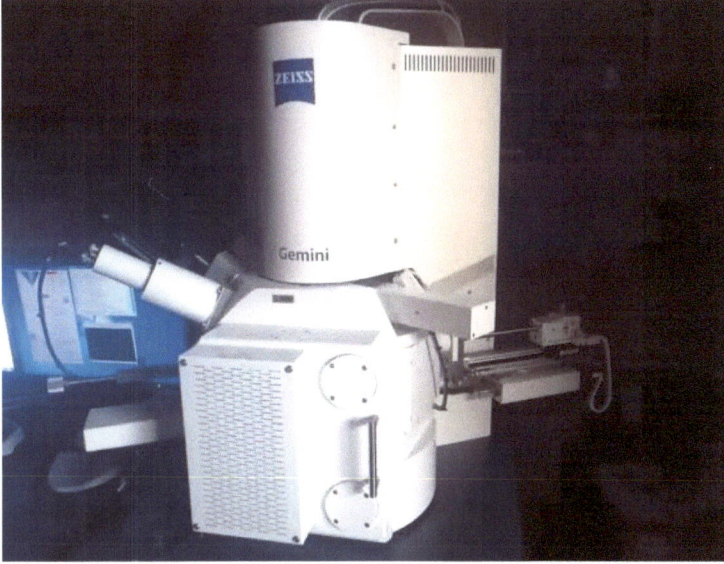

Fig. (17). Electron microscopy (Zeiss Gemini Family) at the University of Limpopo, South Africa (Photo by: Ebrahim Shokoohi).

Fig. (18). SEM photo of a free-living nematode. *Acrobeles iranicus* Shokoohi, Abolafia, and Zad, 2007. (Photo by: Ebrahim Shokoohi).

CONCLUSION

Microscopy plays a crucial role in the field of nematology, as it allows for close examination and identification of nematodes. Accurate measurement of the nematodes is essential for their proper classification, and this is where calibration of the microscope becomes important. Nematode taxonomists rely on precise measurements to distinguish between different species, and therefore, proper calibration of the microscope should always be a top priority. Accurately identifying nematode species is crucial for nematode taxonomists. However, the process of measuring nematodes can vary depending on the group being studied. It is important for taxonomists to have a thorough understanding of the key characteristics that need to be measured to ensure the correct identification of species. Scanning electron microscopy (SEM) and differential interference contrast (DIC) imaging have become indispensable tools in the field of nematology, particularly for the study of taxonomy. These imaging techniques play a crucial role in the analysis of nematodes, especially those with intricate morphological features, such as rhabditids and tylenchids. The detailed images produced by SEM and DIC provide taxonomists with a unique opportunity to study and classify nematodes with greater precision. Therefore, the use of these imaging techniques is considered an indispensable tool for researchers and scientists in the field of nematology.

CHAPTER 6

Molecular Diagnosis

Abstract: Molecular techniques are crucial for research on nematology, and the processing of samples is essential for their long-term storage prior to analysis. This chapter focuses on DNA extraction, various DNA markers, and polymerase change reaction (PCR). To ensure the preservation of nematodes for extended periods, DESS can be utilized. Moreover, the quality of extracted DNA is vital for proper PCR processing. This chapter offers an overview of nematode isolation and the different techniques for DNA extraction. The utilization of molecular markers presents numerous benefits when studying the genetic makeup of nematode populations. Among these markers, sequence characterized amplified region (SCAR) stands out for its exceptional value in identifying nematodes belonging to the *Meloidogyne* species. Amplified fragment length polymorphism (AFLP), on the other hand, is a technique that can be employed to analyze the genetic variability of nematodes. It is important to note that these markers are primarily used for plant-parasitic nematodes, which have a significant impact on the economy. This chapter focuses on the most commonly utilized marker for nematodes. After extracting DNA, the next step involves amplifying the target genes through PCR. Various primers, including rDNA and mtDNA, are available for nematode genes and serve useful taxonomic purposes. However, precise primer design is critical for achieving accurate nematode diagnosis. Both DNA and primers must be of high quality to ensure successful PCR. Eco-friendly standard dyes like SafeView can be employed to visualize PCR products. This chapter offers a comprehensive guide to PCR processing and DNA visualization techniques.

Keywords: Chelex, DNA extraction, DNA marker, Genetic diversity, Primer, Sequencing, Thermocycler, Visualization.

DNA EXTRACTION

Isolation of Nematodes

Nematodes that are extracted from soil or plant tissues can be used for DNA isolation immediately, or they can be preserved in a solution until they can be processed. For the latter, the nematodes can be stored in a 70% ethanol solution for a short period of several months or in a DESS solution for a more extended period. In addition, the nematodes stored in the DESS solution can be used for morphological and molecular studies [41].

DESS Solution

1. Mix 23.265 g of disodium ethylenediaminetera-acetic acid (EDTA = FW: 372.24) for a 250 ml solution (the formula may vary depending on the FW of the EDTA salt). Add 50 ml of deionized water to the EDTA salt and stir (pH = 3-4).

2. Adjust the pH of the EDTA to 8.0 using sodium hydroxide (NaOH). At this stage, about 50 ml of 1M NaOH will be required. By increasing the pH, the EDTA will dissolve gradually, which will be increased by temperature up to 30°C.

3. Using de-ionized water, increase the volume of the solution to 200 ml. Add DMSO to a final volume of 20% DMSO; —for example, 50 ml DMSO for a 250 ml solution.

4. Add NaCl to the solution to be saturated (till it no longer dissolves; the rate of dissolution can be improved by warming to 30°C).

5. Transfer the solution into a container with a lid, and remove any salt crystals.

How to Transfer Nematodes to Ethanol

1. Extract the nematodes from soil or plant tissue

2. Clean a plastic or glass petri dish using a 96 or 70% ethanol solution

3. Add three separated drops of distilled water to the inside of the petri dish

4. Transfer the nematodes to the drops of distilled water one after the other

5. If possible, before the third drop of water, soaking the nematodes in a drop of 70% ethanol would be helpful.

6. After the third drop of water, put the nematodes into a 1.5 ml tube with 20-30 µl of distilled water

7. Keep the tubes that contain the nematodes at 4 °C if the processing can be done in a few days, or at -20°C if the processing will be done only after a week

Note that all steps should be conducted in a sterilized place. The nematodes are ready for DNA extraction.

Genomic DNA Extraction

Isolation of genomic DNA from any type of cell includes three steps [42]:

1. Lysis of cells and extraction of the DNA

2. Dissociation of DNA-protein complexes

3. Extraction of DNA from other macromolecules (carbohydrates, protein, and lipids)

DNA of the nematodes can be isolated manually or by using a DNA extraction kit. Generally, a kit for DNA extraction is more expensive than when the solutions are prepared manually. Moreover, more reactions can be done using manually prepared DNA extraction solutions, while reactions to be done using a DNA extraction kit are limited. Several DNA extraction kits are available, *e.g.*, ClearDetections DNA extraction kit (http://www.cleardetections.com/nematode-dna-extraction) and kits provided by Sigma and Promega companies.

DNA Extraction Using the Chelex Method

Materials: Chelex resin (Sigma sells this as "chelex 100 or chelex 50"), Proteinase K, Sterile water. The schematic view of the process is given in Fig. (**1**).

1. Pprepare a 5% Chelex solution (*e.g.*, for a 10 ml solution)

2. Place a stirrer bar in a 50 ml conical tube held upright in a beaker

3. Place 0.5g Chelex resin into the conical tube

4. Add sterile water to obtain a volume of 10 ml

This Chelex solution can be kept in the fridge for up to 1 month. It is recommended to prepare a fresh work solution for each DNA extraction process [43].

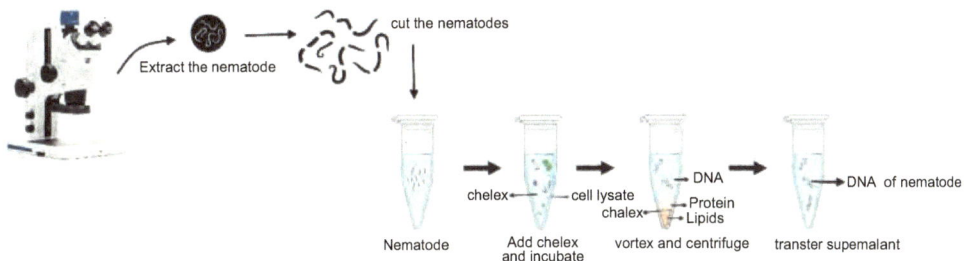

Fig. (1). Overview of DNA extraction of nematodes using Chelex.

Procedure:

1. Crush the nematodes inside the tubes by using a sterilized needle or tip

2. Cut the end of a tip (or smaller tips) to make the opening bigger (Chelex beads are too big to fit otherwise)

3. Mix the Chelex solution using a stir bar (make sure Chelex beads are spinning in the water) and add 20 µl of this to the tubes (make sure Chelex beads are present).

4. Add 1-2 µl of proteinase K solution to each tube (20mg/ml).

5. Centrifuge the tubes and their contents for 30 s

6. Incubate the tubes and their contents at 56 °C for two hours

7. Afterward, incubate the tubes and their contents at 95 °C for 8 min to deactivate proteinase K

8. Cool the tubes and their contents down for a few seconds (if using a thermocycler, bring it down to 40 °C for about 30 sec).

9. Centrifuge the tubes and their contents for about 30 sec

10. Use 4-8 µl of DNA template in the PCR reactions, making sure no beads are included

To make sure the DNA of the nematodes has been extracted during the process mentioned above, NanoDrop can indicate the quantity of DNA extracted.

It should be noted that the PrepGEM method is also used for DNA extraction from individual small nematodes.

Isolation of DNA from Individual Nematodes and their Connected Bacterial DNA

To begin with, isolate the individual nematodes from the culture or soil. Then, wash the living nematodes with solution A (approximately 400 µl) and spin them at 7000 rpm for 2 minutes using a centrifuge. Rinse nematodes from the tube using solution A. Next, put the nematodes in a clean solution A and kill them by heating at 75 °C using a PCR machine block (turn the heated lid off) or dry block. Afterward, wash the nematodes twice using a vortex or centrifugation in solution A (use 0.5 ml tube) and spin at 7000 rpm in the centrifuge for 2 minutes for each wash. Put the nematodes in a small petri dish in solution A and rinse the nematodes from tubes using solution A.

Then, extract the nematode from the original petri dish into a new petri dish containing about 200 µl of solution A. Cut the nematode using a needle, and for each nematode, use a new needle to avoid contamination across the species. Transfer the nematode pieces to a 0.5 ml tube containing 50 µl of tissue and cell lysis solution (Epicentre kit). Express nematodes from the tip and recheck the tube or tip for pieces. Freeze/thaw the tubes on dry ice three or more times. Subsequently, add 1 µl of proteinase K (20 mg/ml). Set the tube at 65 °C for one hour in a PCR machine with a heated lid. Check for the progress of nematode digestion. Continue 65 °C incubation as necessary to complete digestion. Deactivate the enzyme at 95 °C for 10 minutes using a PCR machine.

Add 2 µl of lysozyme (10mg/ml) to each tube and set at 37 °C for 30 minutes. Deactivate the enzyme at 95 °C for 10 minutes using a PCR machine. Next, add 1 µl of proteinase K (20 mg/ml). Set the tube at 65 °C for one hour in a PCR machine with a heated lid. Deactivate the enzyme at 95 °C for 10 minutes using the PCR machine. Add 30 µl of protein precipitation reagent (Epicentre kit). Vortex and centrifuge for 10 minutes. Transfer the supernatant from step 12 to a new 1.5- or 1.7-ml tube. Add 7 µl of the polyacrylic carrier. Add 200 µl of 2-propanol (isopropanol). Mix the contents. Precipitate DNA with overnight incubation in a -20° freezer.

Finally, centrifuge at maximum rpm for 15 minutes to recover the pellet. Wash the pellet twice with 70% ethanol. Dry the pellet and re-suspend in 10-15 µl lambda TE buffer.

Note that Solution A includes 0.05% Triton X-100 in PCR-grade water (12.5 µl in 25 ml Molecular Biology Grade water). Lysozyme solution includes 10 mg/ml in 100 mM Tris-HCl pH 8, and then aliquot and freeze it [43].

DNAzol kit DNA Isolation Protocol for Individual or Pooled Nematodes

To begin, place the recently acquired nematode specimens in a 0.5 ml tube with 100 ul of digestion solution that contains freshly prepared Proteinase K (usually 0.005 g/0.5 ml). Allow the nematodes to digest for at least 0.5 hours at room temperature (Table **1**).

Next, incubate the tube overnight in a 56°C water bath. After incubation, use a 95°C program in a PCR machine for 15 minutes to heat-kill the proteinase. Freeze and thaw the contents of the tube four times.

Table 1. Ingredients for nematode DNA extraction using DNAzol kit.

Ingredients	Amount (µl)
100 mM Tris HCl pH 7.6	200
200 mM NaCl	200
0.5 M EDTA pH 8.0	400
10% Sarkosyl	200
Proteinase K (10 mg/ml)	20
Ultrapure water	980

Centrifuge the sample at a speed of 10000 rpm for 5 minutes. Remove 95 ul of the solution, leaving the bottom 5 ml, and add it to a 1.7 ml tube containing 1 ml of DNAzol isolation reagent.

Add 6 ul of polyacryl carrier solution, ensuring it is well mixed beforehand. Invert the tube 5-6 times to blend the contents. Then, add 0.5 ml of 100% ethanol and mix it again by inverting it 10-12 times.

Store the sample at room temperature for 5 minutes, then use centrifugation at 7000 rpm for 5 minutes to pellet the DNA. Discard the ethanol and wash the DNA twice with 800 ul of 75% ethanol, spinning again if the pellet breaks loose.

Discard the ethanol and remove all with a pipette. Allow visible ethanol to evaporate, but do not let the pellet dry. Finally, resuspend the DNA in 6-20 ul of TE, depending on the number of nematodes you started with (less TE for fewer nematodes to make the DNA more concentrated) [43].

NaOH Digestion of a Single Nematode

Begin by placing one nematode in 20 µl of freshly prepared 0.25 M NaOH. To make this solution, dissolve 1 pellet in 11.6 ml of water each day. Make sure the nematode is fully submerged by spinning the tube. Incubate the tube overnight at 25°C, then heat it for 3 minutes at 99°C. Allow the tube to cool to room temperature and spin it to collect any liquid on the sides. Add 4 µL of 1 M HCl, 10 µL of 0.5 M Tris-HCl (pH = 8.0), and 5 µL of 2% Triton X-100 to the tube. Mix thoroughly and spin the tube again. Warm the tube for 3 minutes at 99°C, then freeze it at -20°C. Warm it up again at 99°C for 3 minutes. Afterward, let the tube cool down to room temperature and check the pH. It should be between 8 and 9 during digestion. Finally, use 1-2 µl of the nematode DNA per 25 µl PCR reaction [44, 45].

Nucleic Acid Extraction using Lysis Buffer I

To prepare the Lysis buffer I, follow these steps:

First, combine 200 mM NaCl, 200 mM Tris-HCl (pH = 8.0), 1% (v/v) β-mercaptoethanol, and 800 µg/ml proteinase K. Remember to add β-mercaptoethanol and proteinase K only shortly before use. Next, pipette the DESS solution into a glass well. Then, transfer the nematode to a PCR tube containing 25 µl of MGW and 25 µl of Lysis buffer I. Afterward, place the PCR tube with the nematode in a thermos shaker at 300 rpm and incubate it for 1 hour and 30 minutes at 65°C. Deactivate the proteinase K by incubating for 5 minutes at 99°C or in a thermocycler. Once DNA extraction is complete, there is no need for DNA clean-up [46].

Finally, you can use the extracted DNA directly or store it at −20°C or below until use.

DNA Extraction from Glycerine-Embedded Nematode Specimens

This method was used with success by the author of this chapter for paratype slides on which *Longidorus artemisiae* were mounted in 1995; *L. goodeyi* prepared them in 1991; two unidentified *Longidorus* spp. from Belgium and Israel were prepared in 1999 and 2000, respectively, and females of *L. leptocephalus* were collected in 1997 and preserved in formalin [47].

1. Open the permanent slide and transfer the nematode carefully to a petri dish half-filled with phosphate-buffered saline (PBS) (PBS: 137 mM NaCl, 2.7mM KCl, 10 mM Na_2HPO_4, 2mM KH_2PO_4).

2. Put the petri dish onto an orbital shaker at low speed.

3. Add fresh PBS after two hours and keep the petri dish overnight.

(Steps 1-3 can be used for the freshly formalin-fixed specimens also)

4. Put the single nematode into 13 µl double distilled water, cut the specimen into 3-5 pieces, and transfer these pieces to an Eppendorf tube.

5. Add 10 µl of lysis buffer (20 mM Tris-HCl (pH=8), 100 mM KCl, 3.0 mM Mg_2Cl, 2.0 mM DDT, 0.9% Tween 20) and 0.1 µl proteinase K (20 mg l-1)).

6. Spin the tube with the nematode-suspended material and store the tube at -70oC for at least 10 min.

7. Incubate the tube and its contents at 65 °C for one hour and then cool it down to 4 °C.

8. Keep the nematode DNA that has been extracted at -70 °C for further analysis

Lysis Buffer Method

1. Transfer 5-10 nematode individuals to a 1.5 ml Eppendorf tube filled with 20 μl of dd-water

2. Incubate the tube in liquid nitrogen for 1 min and then in hot water (T = 80°c) for 5 min

3. Crush the nematode specimens inside the tube with the tip of a needle

4. Incubate the tube with the nematode specimens in liquid nitrogen for 20 s and then in hot water (T = 80°C) for 1 min

5. Centrifuge the tube with its contents for 5 min at 5000 rpm

6. Add the worm lysis buffer and 2 μl of proteinase K

7. Incubate the nematode DNA at 56 °C for two h and then at 95 °C for 8 min

Worm lysis buffer: KCl 1M (50 μl); Tris (pH = 8.3) 1M (10 μl); $MgCl_2$ 1M (2.5 μl); Tween 20 (4.5 μl); dd-water (up to 1 ml).

COMMON MOLECULAR MARKERS

Sequence Characterized Amplified Region (SCAR)

Specific primers were established to develop, using PCR, denoting repetitive sequences' repetitive regions, which are sequence-characterized amplified regions (SCARs) [48]. Specific primers are designed that are 8-10 nucleotides in length. Several root-knot nematode species, including *Meloidogyne arenaria*, *M. chitwoodi*, *M. enterolobii*, *M. fallax*, *M. hapla*, *M. incognita*, and *M. javanica* [48], have been studied using SCAR. The result indicated that these primers have enough sensitivity for diagnosing the species. The SCAR marker has also been used for Soybean cyst nematode (SCN), *H. glycines,* and *H. schachtii* [49]. Their result showed that SCAR is a rapid, reliable, and simple procedure to detect SCN (Fig. **2**).

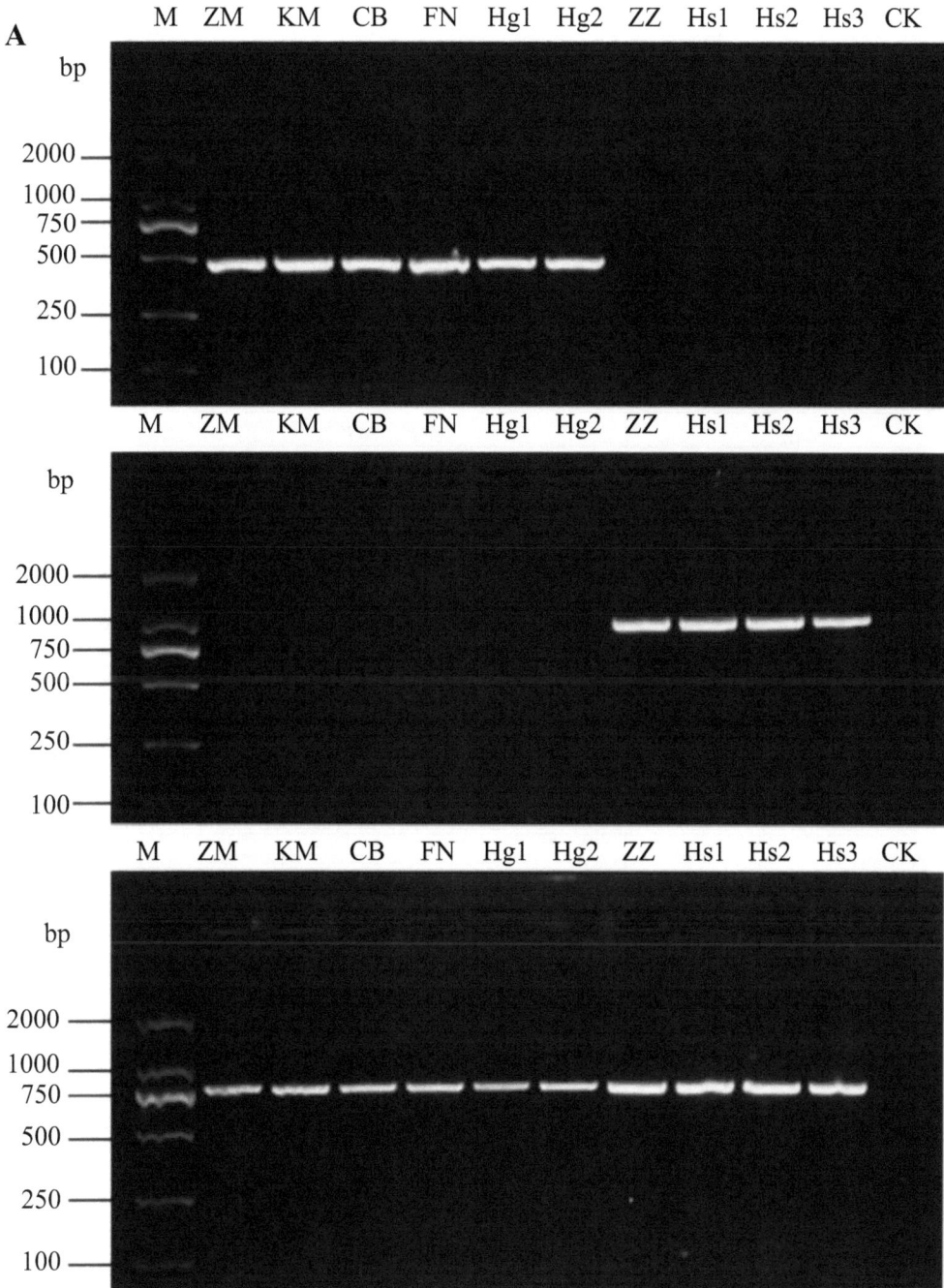

Fig. (2). A SCAR ultraviolet illumination photograph of *Heterodera* from China (adapted from Wu *et al.* [49]).

SCAR primers for the most important plant-parasitic nematodes are given in Table **2**.

Table 2. Primers for SCAR marker to diagnose the most important *Meloidogyne*, *Globodera* **and** *Heterodera* **species [F and R mean Forward and Reverse, respectively].**

Root-knot nematodes (*Meloidogyyne* spp)			
species	code	primer	Reference
M. arabicida	ar-A12F	TCGGCGATAGTACGTATTTAGCG	[50]
-	ar-A12R	TAGTGATTTCGGCGATAGGC	
M. arenaria	Far	TCGGCGATAGAGGTAAATGAC	[51]
-	Rar	TCGGCGATAGACACTACAAACT	
M. chitwoodi	Fc	TGGAGAGCAGCAGGAGAAAGA	[51]
-	Rc	GGTCTGAGTGAGGACAAGAGTA	
M. enterolobii	MK7F	GATCAGAGGCGGGCGCATTGCGA	[52]
-	MK7R	CGAACTCGCTCGAACTCGAC	
M. exigua	ex-D15F	CATCCGTGCTGTAGCTGCGAG	[52]
-	ex-D15R	CTCCGTGGGAAGAAAGACTG	
M. fallax	Ff	CCAAACTATCGTAATGCATTATT	[51]
-	Rf	GGACACAGTAATTCATGAGCTAG	
M. hapla	JMV1	GGATGGCGTGCTTTCAAC	[53, 54]
-	JMV2	TTTCCCCTTATGATGTTTACCC	
-	JMVhapla	AAAAATCCCCTCGAAAAATCCACC	[53]
-	Fh	TGACGGCGGTGAGTGCGA	[51]
-	Rh	TGACGGCGGTACCTCATAG	
M. incognita	MI-F	GTGAGGATTCAGCTCCCCAG	[55]
-	MI-R	ACGAGGAACATACTTCTCCGTCC	
M. izalcoensis	iz-AB2F	GGAAACCCCTAATTAGGATACACT	[50]
-	iz-AB2R	CGCTTGATTTGAGCAGTAGG	
M. javanica	Fjav	GGTGCGCGATTGAACTGAGC	[51]
-	Rjav	CAGGCCCTTCAGTGGAACTATAC	
M. luci	Mlf	ACTCCTGCGACCTCATGGCATTTA	[56]
-	Mlr	ACTCCTGCGAACACAACATTTACT	
Cyst nematodes			
Globodera rostochiensis	-	GCAAGCCCAGCGTCAGCAAC	[57]
-	-	GAACATCAACCTCCTATCGG	

(Table 2) cont.....

G. pallida	-	TGTCCATTCCTCTCCACCAG	[57]
-		CCGCTTCCCCATTGCTTTCG	
Heterodera avenae	HaF1	TGACGAGAACATATGATGGGGATGAT	[58]
-	HaR1	GAGGGGGTGGGAATGAAATGGAT	
H. filipjevi	HfF1	CAGGACGAAACTCATTCAACCAA	[59]
-	HfR1	AGGGCGAACAGGAGAAGATTAGA	
H. glycines	SCNFI	GGACCCTGACCAAAAAGTTTCCGC	[60]
-	SCNRI	GGACCCTGACGAGTTATGGGCCCG	
H. schachtii	OPA06-HsF	GGACCCTGACGACCAGAATA	[61]
-	OPA06-HsR	GACAACACGAAGGAGCGAGC	

Primers (Case Study: *Meloidogyne*)

A set of primers is used for the SCARs study of *Meloidogyne* presented in Table 2.

Procedure:

1. Extract the DNA according to the method described [62]. The Chelex method is recommended since it is fast and easy to do and has a lower cost than a DNA extraction kit.

2. Dilute the primers to 10 pmol concentration

3. Prepare the PCR mixture: 12.5 μl Master Mix, 1 μl forward primer, 1 μl reverse primer, 1-2 μl DNA template, up to 25 μl dd-water

4. Run the PCR for each primer set separately according to the specific program based on each primer's annealing temperature for each species. In this step, you should prepare a PCR mixture for a standard DNA template of each species and control that consists of water only.

5. Run the PCR product on gel electrophoresis and observe it using the gel doc system

Restriction Fragment Length Polymorphism (RFLP)

Another PCR-based method, namely restriction fragment length polymorphism (RFLP), is a technique that shows variations in homologous DNA sequences [63, 64]. Using RFLP, the DNA template is cut into several pieces based on the enzyme used for this method. This method can be used for DNA fingerprinting in forensic investigations, validation of ancestors and parents, and genetic diversity in nematodes and other organisms for various purposes (Fig. **3**).

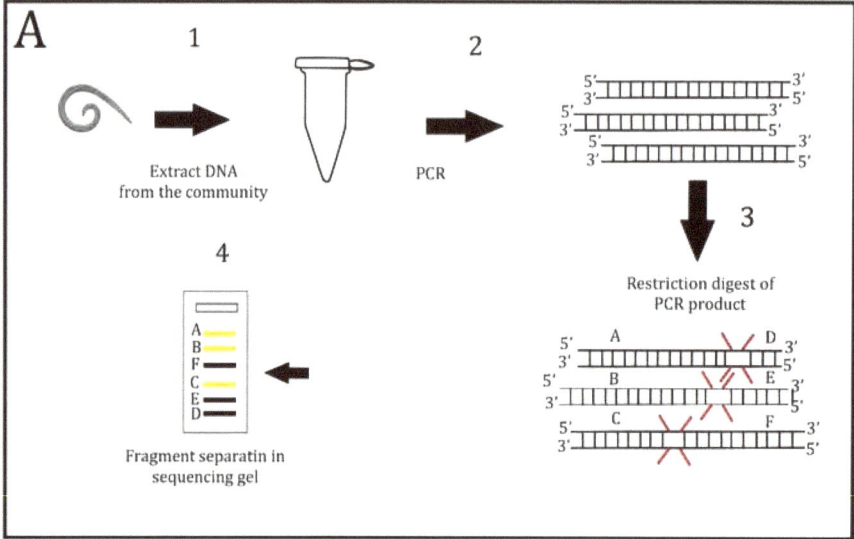

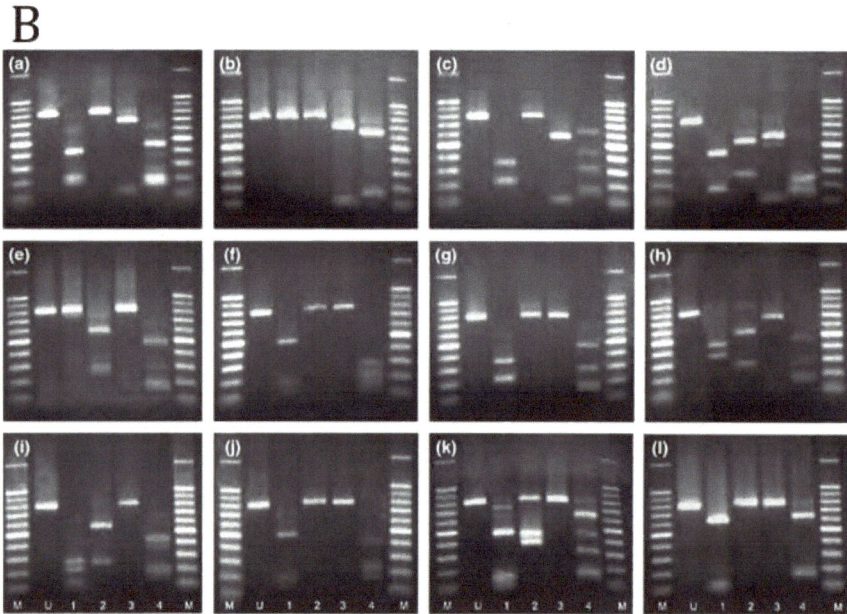

Fig. (3). A schematic plan of results of using the RFLP technique (top) and an ultraviolet illumination photograph of Trichodoridae (bottom) (adapted from Kumari and Subbotin [79]).

The ITS-RFLP technique is a molecular biology method that is highly effective in identifying and differentiating various nematode species and populations. This approach is widely used in nematology research and has proven to be a valuable tool in the study of nematode ecology, population genetics, and diagnosing of nematodes, including *Aphelenchoides* [65], *Bursaphelenchus* [62], cyst forming nematodes [63, 64, 65, 66, 67], *Ditylenchus* [65, 68], *Meloidogyne* [69, 74], *Nacobbus* [70], *Pratylenchus* [71, 72], *Radopholus* [73], *Globodera* species [75], and *Xiphinema* [76]. This method is mainly suited to identify monospecific probes' nematodes. However, this method is not helpful for the identification of mixed-species populations. The application of ITS-RFLP enabled a recent study on genetic variation among different *Heterodera* species, specifically *H. avenae*, *H. filipjevi*, and *H. latipons* [77]. The findings indicated the presence of genetic variation among the different species while noting the absence of such variation within the same population of each species. In addition, ITS-RFLP was used to find out the variation among *D. destructor* [78].

Preparation of DNA Digestion Master Mix

To prepare the solution, take 1 µl of template DNA (1 µg/µL), 2 µl of 10x RE buffer, 1 µl of restriction enzyme, and 16 µl of double distilled water. Note that a restriction enzyme (RE) typically comes with a 10x buffer.

After preparing the solution, set the reaction at the digestion temperature, which is usually 37°C, for 1 hour.

Once the digestion is complete, deactivate it by heating up the mixture to 65°C for 15 minutes. Alternatively, you can add 10mM final concentration EDTA instead of heating.

Now, the digested DNA is ready for further investigation. When using two restriction enzymes simultaneously, it is important to carefully examine the enzyme activities in each buffer before proceeding with the double digestion protocol. This means checking the percentage activity of each enzyme in each buffer to ensure that both enzymes are active in the same buffer. If both enzymes have 100% activity in the same buffer, then you can safely proceed with the double digestion protocol using that buffer. However, if the enzyme activities are not optimal in any of the buffers, it is recommended to determine the optimal buffer to use. This can be done by testing the enzyme activities in different buffers and selecting the buffer that yields the highest enzyme activity for both enzymes. In some cases, buffer incompatibility may cause issues with the double digestion protocol due to the composition or temperature of the buffer. In such cases, sequential digestion may be recommended. This means digesting the DNA with one enzyme first, then purifying the DNA and digesting it with the second enzyme in a different buffer. It is important to note that the digestion solution can be maintained overnight if needed, as long as it is kept at the proper temperature and conditions [64, 80].

Post-digestion Steps

5. First make a 60 ml solution of 2-3% agarose gel. Afterward, the gel must be cooled down.

6. Mix 1 µl of SafeView dye or any other safe dye with 4-6 µl of each PCR product.

7. Fill the gel box with 0.5X TBE, and then pour 4 µl of the ladder into the first and middle wells. Then, pour the PCR products into their respective lanes, which are already mixed with the dye.

8. Use the power supply and set it at 120 V for 30 minutes [64].

9. Once the operation is finished, take the gel for evaluation under the gel doc system to view the polymorphism bands.

For optimal digestion, it is recommended to use a total volume of approximately 10-15 µl [64]. In the protocol presented by Sigma and Promega, this volume goes up to 20 µl. Occasionally, the amplification of DNA through PCR can lead to a feeble product band. The majority of restriction enzymes are derived from *E. coli* bacteria, which flourish at a temperature of 37°C, which is the optimal temperature for their functionality. Yet, there are a handful of restriction enzymes, such as TaqI from *Thermus aquaticus*, which operate best at temperatures that deviate from 37°C.

Amplified Length Fragment Polymorphism (AFLP)

Amplified Length fragment polymorphism (AFLP) is a DNA fingerprinting method that Zabeau and Vos [81] established. This marker relies on PCR of restricted fragments ligated to synthetic adaptors and amplified using primers that relay selective nucleotides at 3' ends. The amplified fragment length polymorphism (AFLP-PCR) method is relatively simple, inexpensive, swift, and consistent in generating high-resolution molecular markers [82]. This procedure uses the grouping of PCR and restriction fragment analysis and supports the recognition of the different DNA regions throughout the genome [83]. DNA fragments range from 60 to 500 base pairs. The advantage of this method is that no prior knowledge of the DNA region sequence is needed. AFLP technique has been used for genetic variation in some nematodes such as *Meloidogyne* species [84, 85], *Heterodera cajani* [86], and *Oesophagostomum bifurcum* [87] (Fig. **4**).

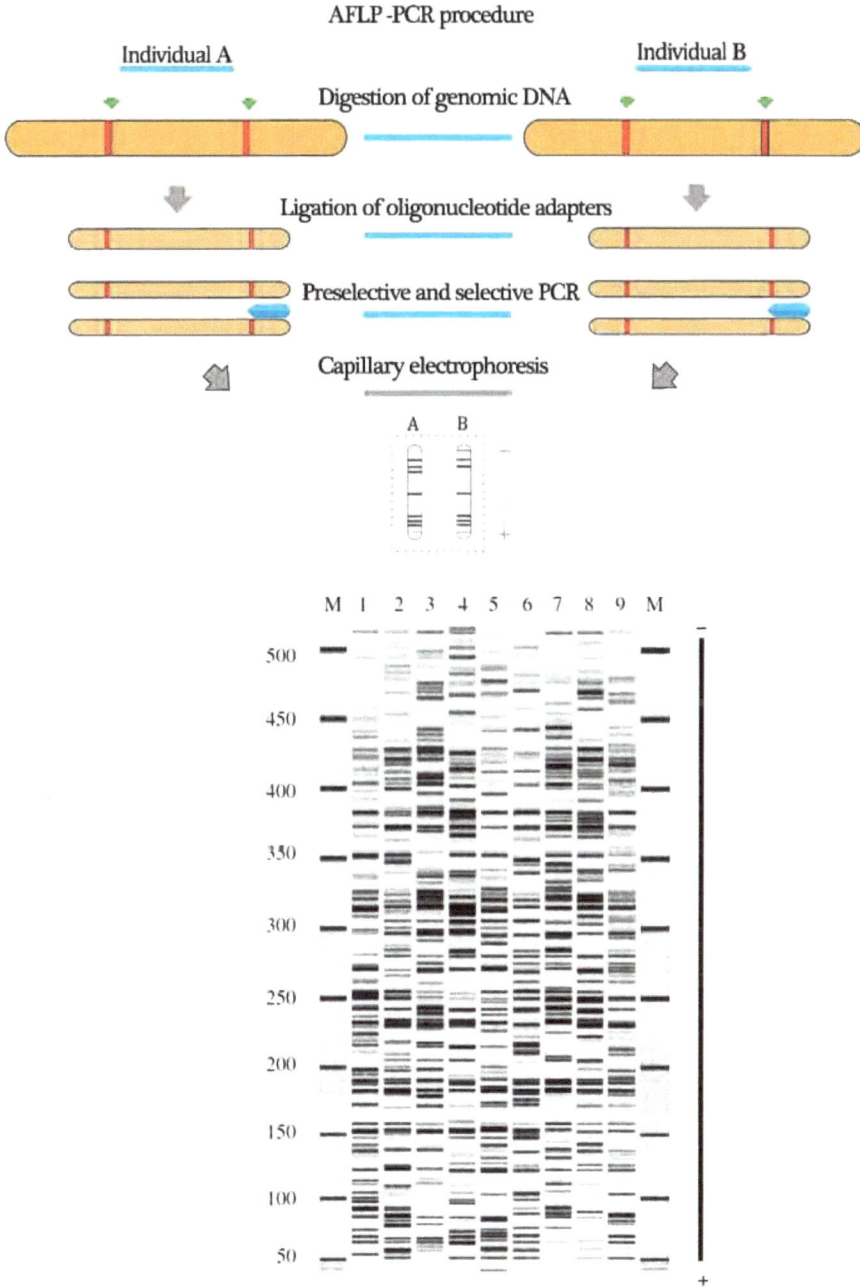

Fig. (4). Schematic plan of AFLP-PCR at the top and an ultraviolet illumination photograph of *Oesophagostomum bifurcum* isolated from humans in Ghana (Lanes 1–9) using the restriction enzyme primer combination HindIII+AG/BglII+AC [87].

When compared to other marker types, AFLP stands out for its remarkable efficiency and ability to provide a wealth of information. This makes it a highly dependable and valuable tool for a range of purposes. However, it does have a drawback in that it produces dominant markers rather than co-dominant ones, and the use of automated systems can result in substantial costs. Rao *et al.* [86] studied variation in Indian populations of pigeon pea cyst nematode, *Heterodera cajani,* using AFLP analysis (Fig. **5**). The AFLP technique has not been in use for over a decade by nematologists, and therefore, the protocol for the practical work is not included in this book.

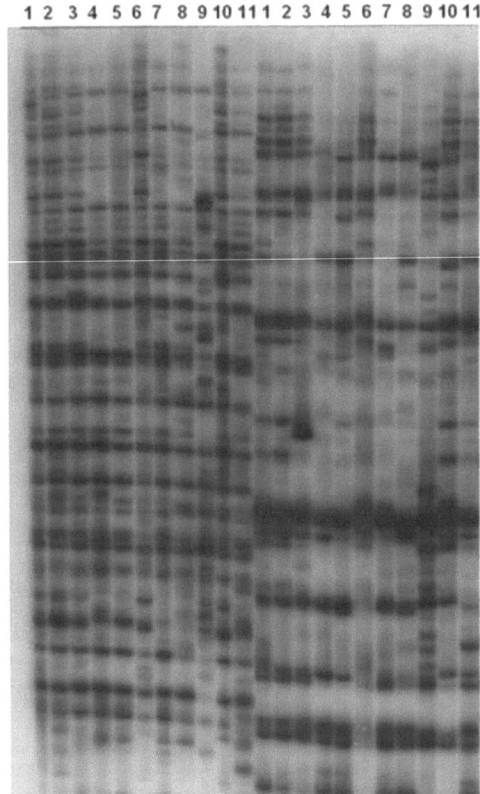

1 2   3 4   5 6 7   8 9 10 11 1 2   3 4   5 6   7 8   9 10 11

Fig. (5). AFLP results for *Heterodera cajani* with *EcoRI* (+AAG) + *MseI* (+CAG) and *EcoRI* (+AAA) + *MseI* (CTA) for different populations from India [86].

Polymerase Chain Reaction

Polymerase chain reaction (PCR) was established as a technique to copy the DNA of live specimens and synthesize large quantities of the target gene by Mullin in 1983. Two considerable improvements, including Taq polymerase and the thermal cycler, have made PCR famous and routinely used today. In 1986, Taq polymerase was isolated from *Thermus aquaticus*, a hot springs bacterium, which was a big improvement in PCR techniques [69]. This technique is now distinguished as one of the most significant innovations of the 20th era. This technique is potent to amplify any organism's specific piece of DNA and can detect a single copy of the target DNA. When preparing a PCR template for amplification, DNA from various sources such as genomic DNA (gDNA), complementary DNA (cDNA), and plasmid DNA can be used. However, the complexity of the DNA affects the optimal input amounts for PCR amplification. For instance, 0.1–1 ng of plasmid DNA is sufficient, whereas 5–50 ng of gDNA may be required for a 50 μL PCR reaction [88].

PCR Processing Comprises Three Steps

1. Denaturation happens at a temperature over 90 °C and separates the double-strand DNA (Fig. **6**).

2. Annealing, which allows the primers to bind with the template DNA of nematodes. For example, the D2-D3 segment of the 28S rDNA of nematodes. At this stage, two strands of DNA are prepared and ready to be copied.

3. Extension, in which a copy of DNA is prepared. At the end of this stage, two identical copies of the original DNA are made. Then, it can be sequenced. The schematic view of PCR steps is shown in the picture below.

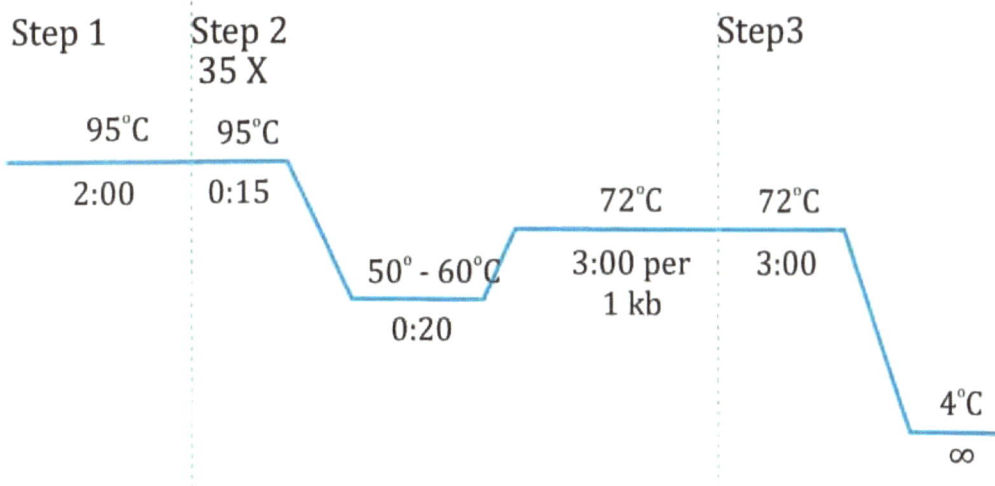

Fig. (6). A schematic plan of the polymerase chain reaction process and the three steps, namely denaturation (Step 1), annealing (Step 2), and extension (Step 3), that the method comprises.

PCR Components

A mixture of PCR includes [89]

dd-Water..volume up to final µl

Final reaction volume ....5 µL to 50 µl

Genomic DNA...10-200 ng

Forward Primer...5-10 pmol

Reverse Primer...5-10 pmol

MgCl$_2$..1.5-5.0 mM (mostly 1.5-2.0 mM)

10x Buffer................1x buffer in final reaction (supplied by the manufacturer)

dNTPs..200 µM each dNTP

Taq DNA polymerase...0.5-1.0 U

These days, Pfu DNA polymerase is also used in the PCR mixture. Pfu DNA polymerase is an enzyme found in the hyperthermophilic archaeon *Pyrococcus furiosus*.

Features of Pfu polymerase: 1) 3'-5' exonuclease activity delivers a low error rate; 2) it is one of the most thermostable DNA polymerases recognized; 3) the lack of additional action means no unwanted 3' overhangs; 4) it is ideal for blunt-ended PCR cloning; 5) its ideal temperature is near 75 °C, and 5) it is 95% active after 1 h of incubation at 98 °C.

Pfu polymerase can be used for 1) high-fidelity PCR and primer-extension reactions, 2) high-fidelity PCR for cloning into blunt-ended vectors, and 3) site-directed mutagenesis.

The PCR master mix (dNTPs, Taq DNA polymerase, and MgCl2) is regularly used in PCR processing since it reduces contamination, the time of the processing process, and pipetting errors. PCR master mix is available in most laboratories and depends on the type of projects the nematologists do. For example, Pfu DNA polymerase is recommended for phylogenetic and standard molecular taxonomy, Taq DNA polymerase and cloning, and whole genome studies. The PCR mixture, when the master mix is used, comprises the following components/agents:

DNA Template

For non-quantitative PCR, 10-100 ng of DNA template is sufficient. The amount of the DNA template strongly influences the success of PCR processing. For plants and animals, genomic DNA up to 500 ng is enough. However, the recommended amount of DNA for nematodes is not defined in terms of ng. Usually, digesting a single nematode in 30 ul of nematode digestion buffer and using 0.5 ul of lysate as template DNA for 16 ul of PCR reaction works for regular ribosomal and mitochondrial DNA processing.

Amplification of the D2-D3 segment of 28S rDNA with 13 ng of DNA template of the target plant-parasitic nematode has been done with success [21]. Success in PCR processing depends on the reaction solutions' quality and the target nematode's DNA template. Depending on the quantity of nematode DNA extracted, 2-8 µl can be used for PCR.

Primers used during PCR: Different primers have been developed for nematological research. Sequence analyses of the 18S rRNA [90, 91, 92, 93, 94, 95, 96, 97, 98, 99, 25, 100], D2–D3 expansion segments of the 28S rRNA [47, 101, 92, 102, 103, 7], ITS rDNA [104], the major sperm protein [106], heat shock Hsp90 [107], actin [108] and expressed sequence tag (EST) [109] genes have been studied for diagnostic and phylogenetic studies by some researchers.

The primers of these regions are used universally and on a routine basis nowadays by most nematology laboratories worldwide. However, the annealing temperature is different for these regions. For example, D2a (5′-ACAAGTACCGTGAGGGAAAGTTG-3′ Tm = 64°C) and D3b (5′-TCGGAAGGAACCAGCTACTA-3′ Tm = 56°C) that are used for amplification of the D2-D3 segment of 28S rDNA having an annealing temperature of 57 °C. The optimum length for the suitable primer should be 18-24 bases, and the optimum GC nucleotide content of the primer should be 40-60%. Generally, the nucleotides G and C should be shared equally along with the primer. The 3′ end of the primer is critical for an efficient extension. The presence of G or C bases within the last five bases from 3′ end of primers (GC clamp) helps foster specific binding at the 3′ end due to G and C bases' robust bonding. Therefore, more than 3 G or C should be avoided in the last five bases at the 3′ end. The ideal number has a stable 5′ and unstable 3′ end. If the 3′ is very stable, this end can bind to a complementary site other than the target, with the 5′ end hanging loose. This can lead to secondary bands. Differences between the melting temperatures (Tm) of the two primers should not be more than 5 °C. For primers with 25 nucleotides, the melting temperature can be computed as follows:

Tm = 4 (G+C) + 2 (A+T)

where G, C, A, and T are the number of those nucleotides in the primer.

For example, concerning the SSU_F_04 (5′-GCTTGTCTCAAAGATTAAGC--3′) and reverse primer SSU_R_26 (5′-CATTCTTGGCAAATGCTTTCG-3′) [103] that have been used for the phylogenetic study on Mononchina [99], the melting temperature for the forward primers is: G = 4, C = 5, A = 6, T = 6

Tm = 4 (4 +5) + 2 (6+6) = 60 and that for the reverse primer is: G = 4, C = 5, A = 4, T = 8

Tm = 4 (4+5) + 2 (8+4) = 60. Preferably selecting the highest temperature is advised for obtaining the best PCR products. PCR processing for the desired primers for nematodes is summarized below in Table **3**.

Table 3. Processing the polymerase chain reaction to amplify nematode deoxyribonucleic acid with the optimal temperature, time, and number of cycles for a high success rate*.

Steps	Temperature °C	Time	Number of cycles
Initial denaturation	94	3 min	1
Denaturation	94	45 s	
Annealing	56	45 s	35
Extension	72	1 min	
Final extension	72	6 min	1

*This program is a case study for SSU_F_04 and SSU_R_26 of 18S rDNA [99].

Quality of DNA Template

To ensure that your DNA is pure, we highly recommend conducting an agarose gel test. This test can easily identify any contamination in the DNA and indicate the presence of purified DNA as a single band on the gel.

It is also crucial to assess the ratio of A260/A280 in your DNA template. A ratio of 1.8 indicates pure DNA, and a ratio of 1.5 to 1.6 is suitable for very clean DNA samples in pure water. A ratio lower than 1.8 may indicate the presence of protein or organic contaminants. However, high-quality results can still be achieved even with ratios below 1.8.

To measure both A260 and A280, an optical spectrometer is used at 260 nm and 280 nm wavelengths, respectively (Fig. 7). A260 measures DNA/RNA concentration, while A280 measures protein concentration. A ratio of A260/A280 greater than 1.8 indicates minimal protein contamination in a DNA/RNA sample.

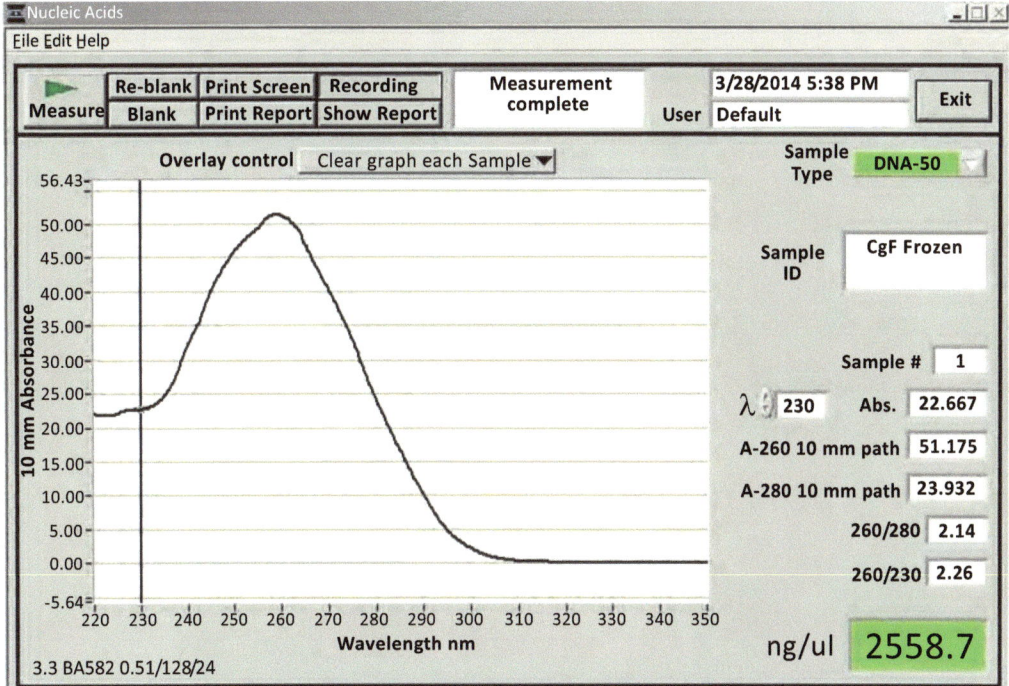

Fig. (7). Amount of genomic DNA and A260/A280 ratio of *Mesorhabditis* sp. [21].

Evaluating the PCR Products

Four microliters of each DNA template and ladder should be mixed with the Gel-Red or safe-view before electrophoresis. This solution is poured into the wells, and the electrophoresis is run at 120 V for 30 min. However, a lower amount of DNA makes a faint band. After the electrophoresis is completed, the gel's DNA template can be stained to make the amplified region visible using a Gel Doc system. When the desired gene is amplified successfully, the PCR product must be stored at -20 °C until all samples have been prepared and labeled to send to a company for sequencing (Fig. **8**).

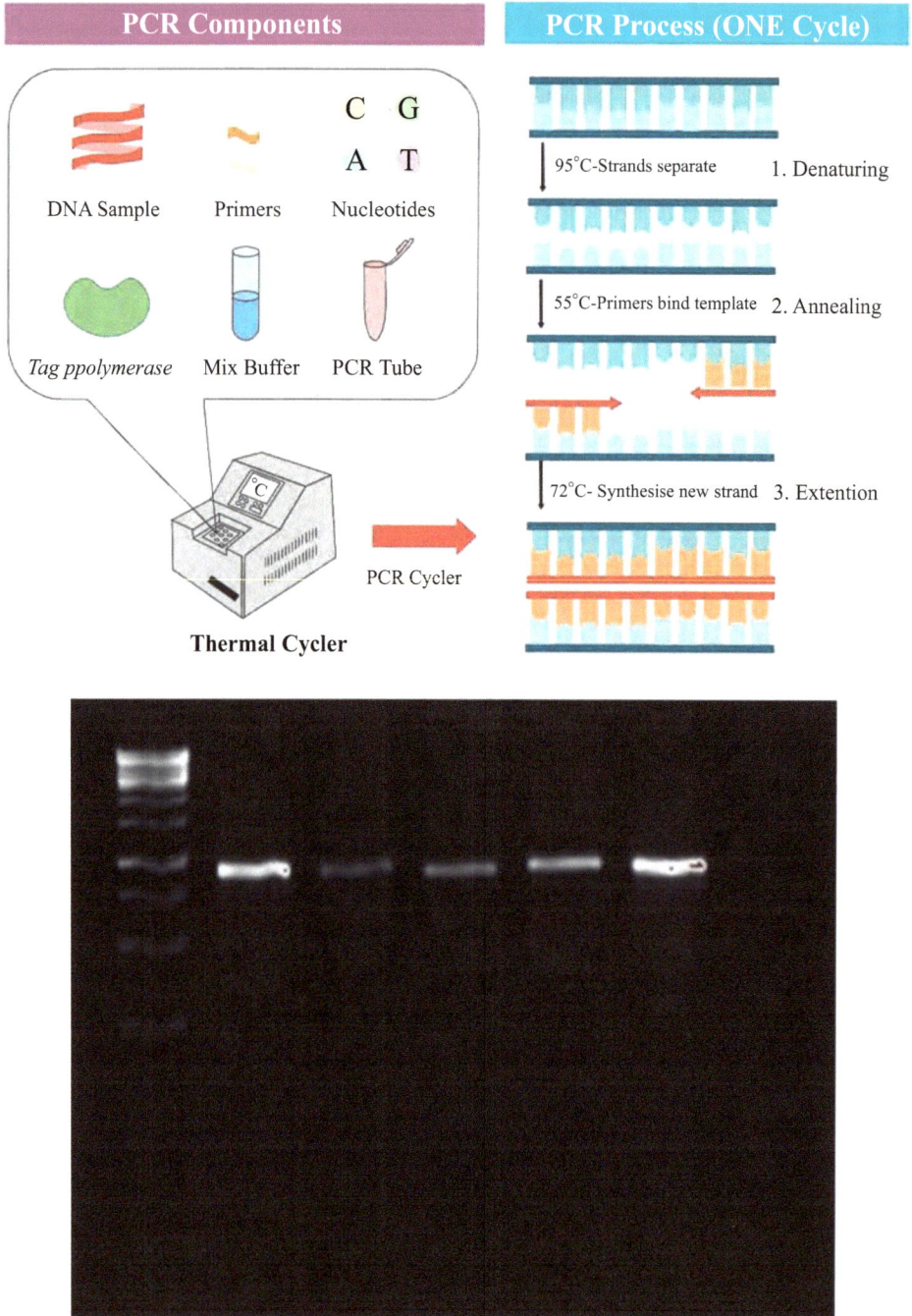

Fig. (8). A schematic view of the polymerase chain reaction process and a gel that contains deoxyribonucleic acid of *Meloidogyne* [105]. (Photo by: Ebrahim Shokoohi)

CONCLUSION

DNA extraction is a vital process in the field of nematology, as it is the foundation of numerous molecular studies. The quality of extracted DNA has a significant impact on the reliability and accuracy of the results obtained from these studies. There are various DNA extraction methods available that can be employed to investigate the phylogeny and molecular taxonomy of nematodes. However, the selection of the method largely depends on the resources available. Although the Chelex method is cost-effective, it may not always provide satisfactory results. On the other hand, using a proper DNA extraction kit can yield high-quality DNA, which is essential for successful molecular studies. Therefore, it is crucial to choose the right DNA extraction method to ensure the accuracy and reliability of the results. Molecular markers are powerful tools that enable researchers to study the genetic diversity of nematodes at a molecular level. Specifically, molecular markers are short fragments of DNA that can be used to track genetic variations and identify specific genes or genetic sequences. The choice of marker and method of analysis is dependent on the feeding habits of the nematodes being studied. For instance, markers that are commonly used to study plant-parasitic nematodes are not always effective for studying free-living or animal-parasitic nematodes. Plant-parasitic nematodes are of particular interest to researchers due to their significant economic impact on agriculture. In order to reveal the genetic diversity of these nematodes, it is necessary to select a suitable molecular marker that can accurately capture the complexity of their genetic makeup. By analyzing the genetic diversity of nematodes, researchers can gain valuable insights into the mechanisms of evolution and the ecological relationships between nematodes and their hosts. Polymerase chain reaction (PCR) is a widely used technique in molecular biology that enables the amplification of specific target genes from nematodes. This technique is crucial for a variety of research studies, as it allows researchers to generate large amounts of DNA from small samples. To perform PCR, a suitable primer and functional master mix are required. The primer serves as a starting point for the amplification process, while the master mix contains all of the necessary components for the reaction to occur. With the help of PCR, researchers can study the genetic makeup of nematodes in greater detail, leading to a better understanding of these important organisms.

CHAPTER 7

Basic Bioinformatics in Nematology

Abstract: Bioinformatics is the interdisciplinary study of nucleic acid and protein sequences, which has proven especially useful for genomics, gene expression, and nematode diagnosis. Quality control of these sequences is essential, and bioinformatics plays a crucial role in their processing for phylogenetic analysis. Skilled analysis is particularly vital for molecular ecology and environmental DNA analysis, especially when working with next-generation sequences. Furthermore, the utilization of an online database and specialized software designed for the identification of nematode species serves as valuable resources for a proper diagnosis of nematode-related issues. In this chapter, we will explore the bioinformatics of Sanger sequencing, with an emphasis on the phylogenetic study of nematodes and online species identification of nematode species.

Keywords: Bioinformatics, DNA, Nematodes, Phylogeny, Sanger sequencing.

GENERAL INFORMATION

A massive amount of data mainly related to molecular analysis indicates the challenges in computing in biological sciences. Bioinformatics is a field of study that leverages computational methods to analyze large datasets associated with biomolecules. It is widely recognized as a subfield of molecular biology and encompasses a range of subjects, including structural biology, genomics, gene expression, DNA barcoding, and phylogenetics.

The main scientific products of Nematology nowadays focus on DNA barcoding to identify the species and phylogenetic study. For this reason, the PCR products can be sequenced by different technologies. The DNA sequencing method, commonly referred to as Sanger sequencing or the chain termination approach, was introduced by Sanger and colleagues in 1977. This technique entails incorporating chain-terminating dideoxynucleotides (ddNTPs) selectively into DNA polymerase during in vitro DNA replication.

Nowadays, a large set of nucleotide data can be obtained by New Generation Sequencing or NGS [106]. A study of the organisms by using NGS can be done on different platforms [106]. DNA divides into some parts in the 454 Roche pyrosequencing method, and then sequencing of 400-600 megabases of DNA

Ebrahim Shokoohi

during the 10-hour running will be obtained. In Illumina sequencing, DNA is first cut, and then two adapters ligate to the end of each. This method yields millions of reads from the original DNA. Illumina is the most usable technique in the field of next-generation sequencing. In Nematological study, the nematode community is mainly studied by the pyrosequencing method of the Roche 454 platform [107, 108, 109, 110, 111]. NGS sequencing is beneficial for understanding the diversity of the nematodes in the related community [106]. An NGS study of nematodes by combining two genes, including SSU and LSU, leads to precisely identifying 97% of the species [107].

CHECKING THE SEQUENCES

Sequencing technologies are imperfect, and quality control is necessary to confirm that the data used for the downstream study is not compromised of low-quality sequences, sequence artifacts, or sequence contamination that might lead to erroneous conclusions. The simplest way of quality control is by looking at summary statistics of the data. Different programs can yield those statistics.

The first step in checking the quality of the sequences is to open and observe the raw data and analyzed files. For this purpose, open the "*.ab1" file (the raw chromatogram trace file) using the "BioEdit" or "Chromatogram" program.

In the chromatogram presented for *Butlerius* (Superfamily: Diplogasteroidea) (Fig. **1**), some features, including well-formed and distinct single-colored peaks, separated peaks, and lacking background signals, are visible. This sequence quality is considered successful. The successful result could be due to the appropriate concentration of DNA template and primer, excellent purity of DNA, and optimum primer design.

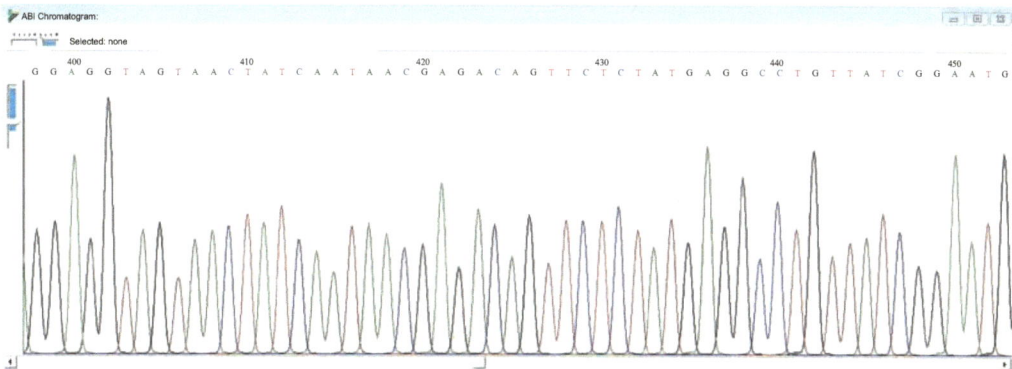

Fig. (1). Chromatogram of a normal rDNA sequencing read of *Butelrius butleri* [112].

In the chromatogram presented for *Helicotylenchus* (family: Hoplolaimidae) (Fig. **2**), some features, including the absence of clearly defined peaks in raw and analyzed data, the presence of excess dye peaks, and a high ratio of noise, are visible. This sequence quality is considered a failed one. The failed result could be due to an inadequate amount or poor quality of DNA template or the primers not binding well.

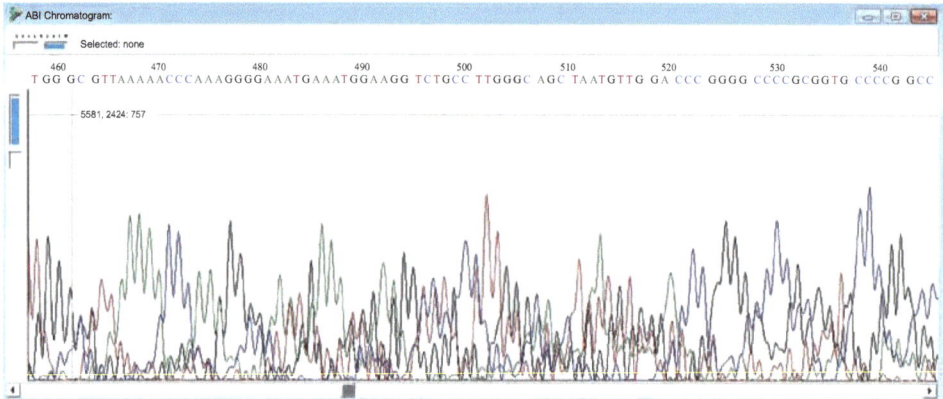

Fig. (2). Chromatogram of a failed rDNA sequencing read of *Helicotylenchus* sp. [21].

In the chromatogram presented for *Helicotylenchus* (family: Hoplolaimidae) (Fig. **3**), some features, including the unclean pick of the sequencing, multiple peaks with the same height or differing heights, overlapping one another, and artifacts beneath the peaks are visible. This sequence quality is considered a failed one. The failed result could be due to a contaminated DNA template or primer, poor DNA template and primer quality, or multiple priming.

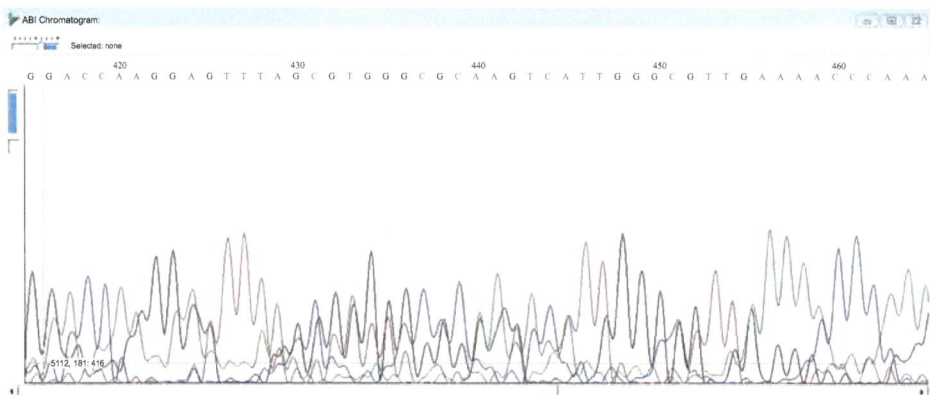

Fig. (3). Chromatogram of a multiple sequence signal of rDNA sequencing read of *Helicotylenchus* sp. [21].

In the chromatogram presented for *Helicotylenchus* (family: Hoplolaimidae) (Fig. **4**), some features, including the mixing of different sequencing peaks, are visible. This sequence quality is considered a failed one. The failed result could be due to unspecific binding due to a mixture of various primer products with differing lengths. To solve this problem, isolate new DNA from a single nematode and re-sequence. On the other hand, optimizing the PCR if multiple bands appear after checking the PCR template on an agarose gel, checking the quality of the primer, and re-designing the primer would be helpful.

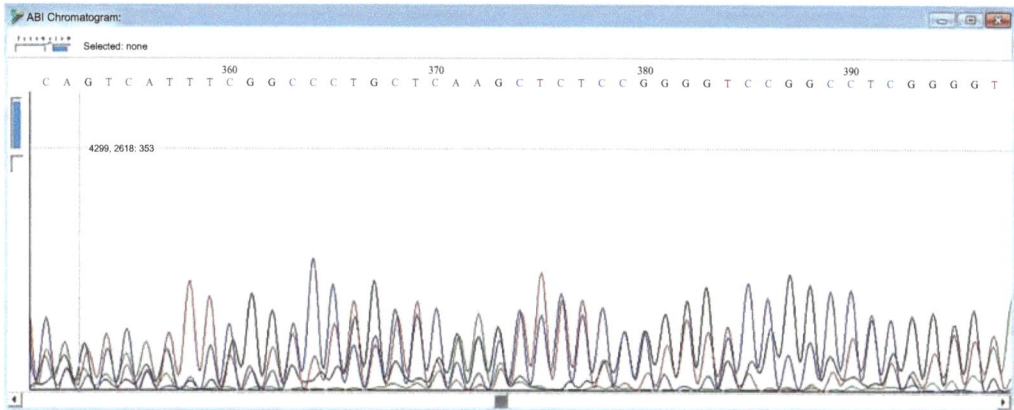

Fig. (4). Chromatogram of a mixing of the peaks of rDNA sequencing read of *Helicotylenchus* sp. [21].

In the chromatogram presented for *Acrobeloides* (family: Cephalobidae) (Fig. **5**), some features, including shallow signals of the peaks in the raw data, are visible. On the other hand, the base calls fade off before the end of the reading. This issue may be due to impure DNA (as we extracted from 70% ethanol) and the low quantity of the DNA template. To solve this problem, re-isolating the nematode with a suitable amount of DNA and removing the inhibitors such as ethanol could be helpful.

In the chromatogram presented for *Acrobeloides* (family: Cephalobidae) (Fig. **6**), some features, including very high peaks followed by sudden stop reading in the raw data, are visible. This issue may be because of too little DNA, too much primer, or impurified DNA template. To solve this problem, checking the concentration of the template DNA using a combination of gel electrophoresis and spectrophotometer readings, checking the concentration of the oligonucleotide primer, and preparing a clean template of DNA can be helpful.

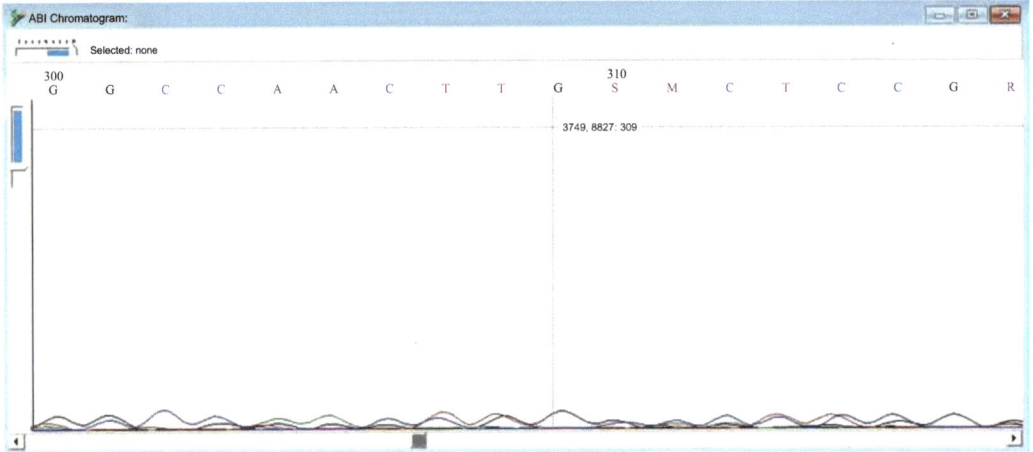

Fig. (5). Chromatogram of a low signal of the peaks of rDNA sequencing read of *Acrobeloides* sp. [21].

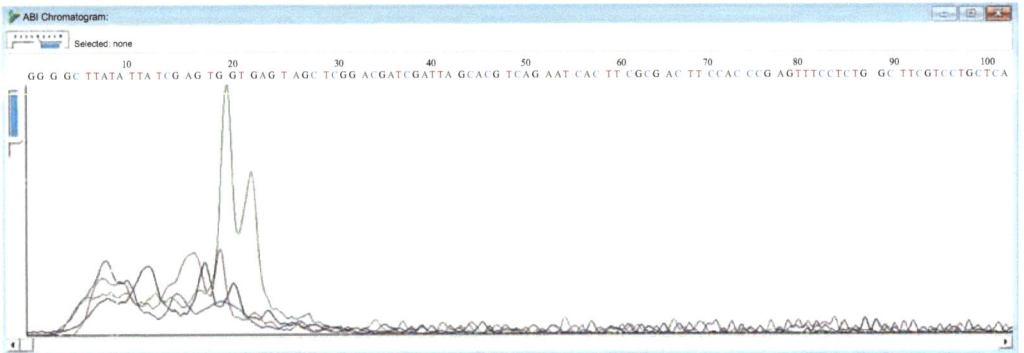

Fig. (6). Chromatogram of a sudden stop reading of rDNA sequencing read of *Acrobeloides* sp. [21].

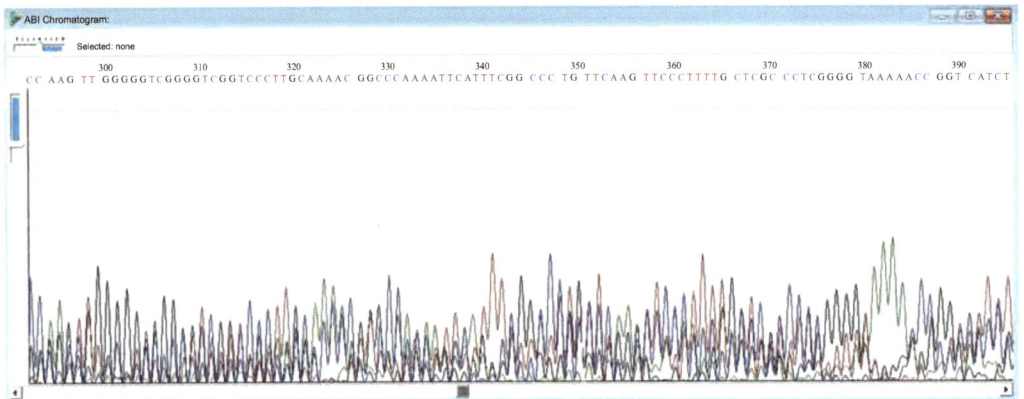

Fig. (7). Chromatogram of mixing peaks of rDNA sequencing read of *Helicotylenchus* sp. showing degraded primers [21].

In the chromatogram presented for *Helicotylenchus* (family: Hoplolaimidae) (Fig. 7), some features, including the mixing of different peaks and low signals of the peaks in the raw data, are visible. This issue is caused by unspecific binding due to a mixture of various primer products with differing lengths. To solve this problem, isolating new DNA from a single pure colony and re-sequencing, optimizing PCR conditions if multiple bands appear after checking the PCR template on an agarose gel, checking the primers for purity, and, if necessary, re-designing the primer can be helpful.

BLAST OF THE SEQUENCES

In bioinformatics, Blast is a "Basic Local Alignment Search Tool" for the comparison of primary biological information, such as DNA sequences. This tool is based on an algorithm primarily used for searching the sequences.

The BLAST algorithm is a heuristic program, which means that it relies on some smart shortcuts to perform the search faster. BLAST makes local alignment. The local alignment approach investigates DNA and protein for their similarity [113]. The word size allowed for the blast is typically 3 for proteins and 11 for nucleotides. The sequence of a nematode should be in FASTA format, as indicated for *Pseudacrobeles macrocystis*, a South African isolate (Fig. **8**).

>Pseudacrobeles_macrocystis

GGTTAGGTACAGACGCGCTGTGGTGTTGCCGGGTTGGCGCTTGGGTGAGTCGCATGCGACAAGTCCAGTTGTT
CGATCTTGGTGGCGTTATAGTGCCCTAGCTGATCCTCGGTGTAAAAGTTGGTCATCTCTCCGACCCGTCTTGAAAC
ACGGACCAAGGAGTCTAGCGTATGCGCGAGTCATTGGGCGGAAAACCCATAGGCGTAATGAAAGTAAAGGTCT
CTTTTCGGAGGCTGATATGCGATCCGTCGTGCCACGGTGCGGCGGAGCAGCATAGCCCCATCTTGACTGCTTGC
AGTGGGGTGGAGGTAGAGCGTATTCGCTGGTACCCGAAAGATGGTGAACTATGCCTGAGCAGGATGAAGCCAG
AGGAAACTCTGGTGGAGGTCCGTAACGCTTCTGACGTGCAAATCGATCGTCTGACTTGGGTAT

Fig. (8). FASTA format of a part of *Pseudacrobeles macrocystis* 28S rDNA sequence from South Africa.

Once you perform the Nblast search on NCBI, a webpage will display the BLAST report. The report kicks off with a succinct overview of the BLAST version utilized, database specifications, and the search query sequence. The standard BLAST report consists of three key components: a graphical summary (as depicted in Fig. **9**), a list of blast hits, and their respective alignments. To make sense of our Nblast search results, we will analyze each of these sections.

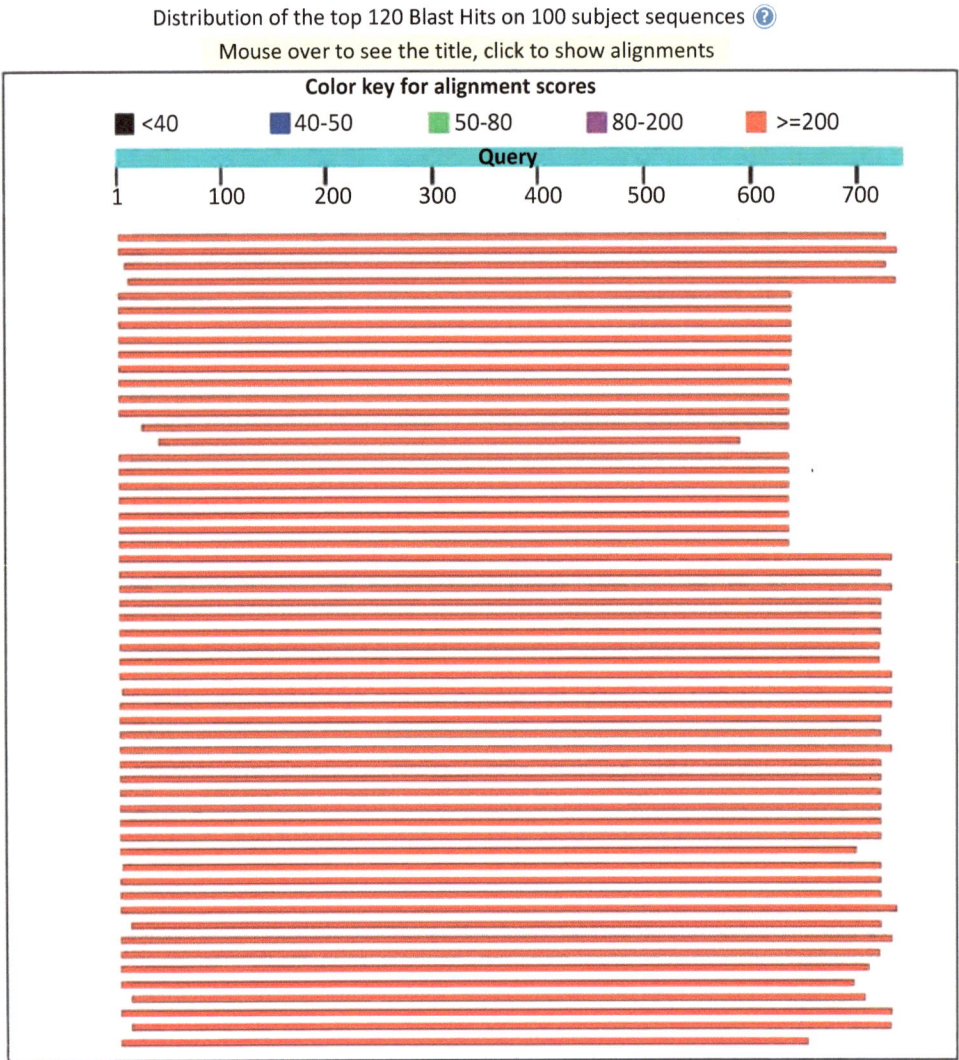

Fig. (9). Graphical view of Nblast result of a 28S rDNA sequence of *P. macrocystis* from South Africa.

Displayed at the top of the figure is the query sequence represented by a green bar, followed by a series of aligned database hits underneath it (Fig. **10**). The first hit displayed is the sequence most closely related to the query. In this particular instance, there are several database hits with impressive scores that align with the majority of the query sequences. The red bar represents the actual *Pseuacrobeles*.

Descriptions	Graphic Summary	Alignments	Taxonomy

Sequences producing significant alignments   Download ⌄   Select columns ⌄   Show 100 ⌄ ❓

☑ select all *100 sequences selected*   GenBank   Graphics   Distance tree of results   MSA Viewer

	Description	Scientific Name	Max Score	Total Score	Query Cover	E value	Per. Ident	Acc. Len	Accession
☑	Pseudacrobeles macrocystis isolate ULps4 large subunit ribosomal RNA gene, partial sequence	Pseudacrobeles...	780	780	100%	0.0	100.00%	432	MW301158.1
☑	uncultured fungus genomic DNA sequence contains 18S rRNA gene, ITS1, 5.8S rRNA gene, ITS2, 28S rRNA g...	uncultured fungus	609	609	100%	2e-169	91.44%	1833	OU940686.1
☑	Eucephalobus striatus isolate 5974 28S large subunit ribosomal RNA gene partial sequence	Eucephalobus st...	609	609	100%	2e-169	91.44%	1038	HM439769.1
☑	Pseudacrobeles sp. OH-2016 28S ribosomal RNA gene, partial sequence	Pseudacrobeles ...	600	600	100%	1e-166	90.74%	3076	KU180684.1
☑	Pseudacrobeles sp. JB-85 28S ribosomal RNA gene, partial sequence	Pseudacrobeles ...	600	600	100%	1e-166	90.74%	1015	DQ145654.1
☑	Acrobeloides sp. IZ-001 28S ribosomal RNA gene, partial sequence	Acrobeloides sp...	575	575	100%	5e-159	90.09%	986	DQ903085.1
☑	Pseudacrobeles bostromi isolate 5612 28S large subunit ribosomal RNA gene, partial sequenc	Pseudacrobeles ...	572	572	100%	6e-158	89.12%	1043	HM439772.1
☑	Eucephalobus mucronatus isolate 5966 28S large subunit ribosomal RNA gene, partial sequence	Eucephalobus m...	572	572	100%	6e-158	89.61%	1044	HM439767.1
☑	Cephalobus sp. CR-2010 voucher 5CR6D9 28S large subunit ribosomal RNA gene, partial sequence	Cephalobus sp...	572	572	100%	6e-158	90.14%	723	HM055399.1
☑	Tricirronema trifilum isolate N690 28S ribosomal RNA gene, partial sequence	Tricirronema trifil...	569	569	100%	2e-157	88.89%	997	GU062818.1
☑	Zeldia punctata 28S ribosomal RNA gene, partial sequence	Zeldia punctata	567	567	100%	7e-157	89.38%	3032	KU180685.1
☑	Zeldia punctata isolate SA large subunit ribosomal RNA gene, partial sequence	Zeldia punctata	567	567	100%	7e-157	89.38%	711	MZ254892.1
☑	Acrobeloides sp. CR-2010 voucher 2CR13D9 28S large subunit ribosomal RNA gene, partial sequence	Acrobeloides sp...	567	567	100%	7e-157	89.86%	735	HM055396.1
☑	Zeldia punctata strain PDL3 28S ribosomal RNA gene, partial sequence	Zeldia punctata	567	567	100%	7e-157	89.38%	3439	EU195988.1
☑	Zeldia punctata large subunit ribosomal RNA gene, partial sequence	Zeldia punctata	567	567	100%	7e-157	89.38%	729	AF147070.1
☑	Zeldia punctata 28S ribosomal RNA gene, partial sequence	Zeldia punctata	567	567	100%	7e-157	89.38%	1030	DQ145662.1
☑	Acrobeloides sp. CR-2010 voucher 5CR3D9 28S large subunit ribosomal RNA gene, partial sequence	Acrobeloides sp...	564	564	100%	9e-156	89.63%	744	HM055398.1
☑	Acrobeloides sp. CR-2010 voucher 3CR1D9 28S large subunit ribosomal RNA gene, partial sequence	Acrobeloides sp...	564	564	100%	9e-156	89.38%	747	HM055397.1
☑	Zeldia punctata isolate HN3 large subunit ribosomal RNA gene, partial sequence	Zeldia punctata	563	563	100%	3e-155	89.15%	659	OM280054.1
☑	Nothacrobeles abolafiai 28S ribosomal RNA gene, partial sequence	Nothacrobeles a...	563	563	100%	3e-155	89.15%	761	KC182515.1
☑	Acrobeloides sp. FHD001 large subunit ribosomal RNA gene partial sequence	Acrobeloides sp...	563	563	100%	3e-155	89.63%	784	MW327029.1
☑	Pseudacrobeles sp. JB-56 28S ribosomal RNA gene, partial sequence	Pseudacrobeles ...	561	561	100%	1e-154	89.12%	1034	DQ145653.1
☑	Pseudacrobeles variabilis large subunit ribosomal RNA gene, partial sequence	Pseudacrobeles ...	561	561	100%	1e-154	89.12%	1035	AF143368.1

Fig. (10). Tubular display of Nblast hits of *P. macrocystis* from South Africa.

BLAST uses information from the database record's definition line to populate the first column. However, due to limited space, some reports may show truncated descriptions.

The second column is "Max score". Max score is the utmost alignment score (bit-score) between the query sequence (*P. macrocystis* from South Africa) and the database sequence fragment (Other *P. macrocystis*). The Max scores for the first and the second rows are 780 and 600, respectively. This means that the *P. macrocystis* from South Africa has more similarity with the MW301158 in the first row.

The final score is computed by summing up the alignment scores of matching segments in the database sequence with the query sequence across all segments. It is worth mentioning that the maximum score may not always be the total score since distinct portions of the database sequence may correspond to different regions of the query sequence. However, in many cases, this score is related to the Max score as visible in the dataset for *P. macrocystis*.

The degree of overlap between query and subject sequences can be gauged by measuring the query coverage, which represents the percentage of the query sequence that matches the subject sequence. For *P. macrocystis*, the top result shows a query coverage of 100%. The E-value denotes the number of expected chance alignments with a specific score or better. It serves as the default sorting metric, indicating the number of sequences that closely match the query sequence and would be expected to appear in a database search. E-values below 1 suggest a high probability of finding a good match, whereas an E-value of 0.01 indicates a 1% chance of finding a suitable match. A lower E-value signifies a better match, and an E-value of 0.0 implies a 0.0% chance of locating the match in a database of random sequences. The E-value aids in assessing the quality of the sequence. An E-value of 1e-50 or smaller suggests a low number of hits with high sequence quality. An E-value of 0.01 indicates that hits smaller than 0.01 are still appropriate for homology matches. An E-value smaller than 10 includes hits that may not be significant but provide an idea of potential relationships. Identity refers to the sequence block with the highest percentage of matching bases. In the present dataset, *P. macrocystis* exhibits a maximum identity of 100%, but this does not necessarily imply that the sequences are identical morphologically. The last column of the table features hypertext accession numbers of the hits in the database. You can follow these numbers to the RefSeq records and gain further insights into these sequences.

ALIGNMENT AND PHYLOGENETIC ANALYSIS

The scientific field of phylogenetics delves into the study of phylogeny, which is a subfield of systematics encompassing taxonomy. Taxonomy is primarily concerned with identifying and classifying different organisms. To infer the relationships between various taxa or sequences and their hypothetical common ancestors, phylogenetic trees are utilized [114, 115]. Currently, most phylogenetic studies rely on molecular data, such as DNA (rDNA, mtDNA) or protein sequences, to create phylogenetic trees. These trees aid in estimating the relationships between different species represented by the sequences. Scientists utilize various software tools to expand on the tree topology of the sequences to determine the evolutionary pathway of organisms. One such software tool is MEGA, which is also known as molecular evolutionary genetics analysis. This computer software performs statistical molecular evolution analysis and constructs phylogenetic trees.

WHAT SEQUENCES SHOULD BE SELECTED?

Prior to developing genomic data, the morphological data was used for species delimitation. Nowadays, phylogenetic analysis is based on genetic features, including nucleotide or protein sequence data, which is useful for various studies. It should be verified that the sequences planned to construct a phylogenetic tree are homologous. For this purpose, we can use BLAST, which is implemented in NCBI. The BLAST tool, which can be used for searching for similar sequences, is available online *via* the EBI or NCBI websites. It should be highlighted that similarity (indicating two sequences are correlated by divergent evolution of a common ancestor) does not indicate homology (similar structures in two or more different species that derive from a common ancestor of the species) because of the possibility of homoplasy (A homoplasy is a trait that is present in a group of species, but not in their common ancestor). The BLAST search procedure includes guidelines for determining homologous sequences based on similarity measures. Next is the multiple-sequence alignment, which assists in elaborating the phylogenetic tree.

ALIGNMENT USING MEGA

1. Select "align" and open the "edit/build alignment" or open the already created text file.

2. Choose "create a new alignment" and press "OK".

3. Choose "DNA". (or Protein; here, we selected DNA as we worked on)

4. Select all and copy your sequences as FASTA files from where they were deposited in your PC.

5. Paste the sequences chosen into the new sheet opened in the MEGA window.

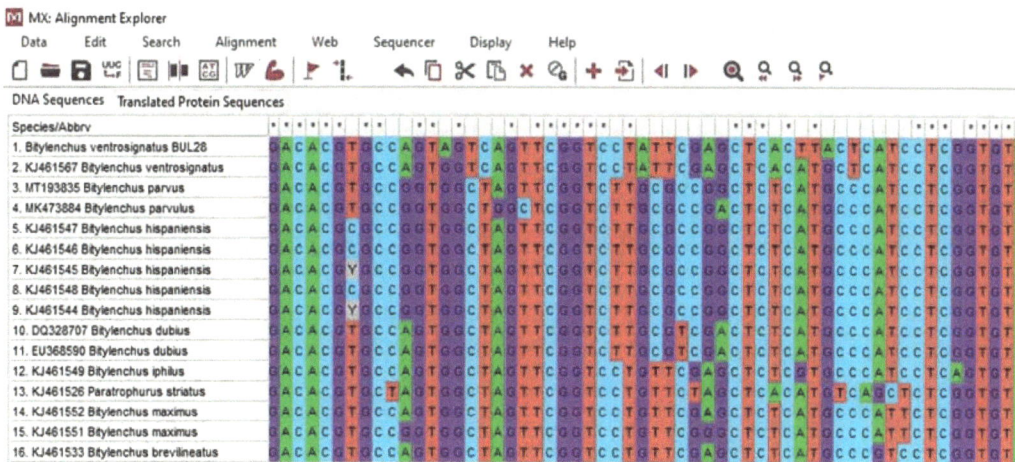

Fig. (11). Multiple alignments of the 28S rDNA of *Bitylenchus* populations using MEGA11.

6. Go to the "Alignment" menu and select "Align by ClustalW".

7. Save the output as "file name.mas", *e.g.* "Bitylenchus.mas".

COMPUTING PAIRWISE DISTANCE

1. Select "Distance" and then "Compute Pairwise command".

2. Use the "Variance estimated method" as bootstrap and justify for 2000 (this can be optional).

3. Use the "Model" as "maximum composite likelihood".

4. Leave the rest as software default.

5. Press "compute" to estimate genetic pairwise distance (Fig. **12**).

MX: Pairwise Distances (gdddd.meg)

File   Display   Average   Caption   Help

	1	2	3	4	5	6	7	8	9	10	11	12	13	14	15	16
1. Bitylenchus ventrosignatus BUL28		0.010	0.033	0.036	0.031	0.041	0.031	0.031	0.032	0.033	0.033	0.027	0.030	0.026	0.026	0.029
2. KJ461567 Bitylenchus ventrosignatus	0.045		0.032	0.034	0.029	0.038	0.029	0.028	0.031	0.032	0.032	0.029	0.031	0.026	0.027	0.029
3. MT193835 Bitylenchus parvus	0.186	0.183		0.007	0.004	0.007	0.003	0.004	0.003	0.009	0.009	0.026	0.027	0.022	0.022	0.023
4. MK473884 Bitylenchus parvulus	0.198	0.193	0.023		0.007	0.007	0.007	0.007	0.007	0.009	0.009	0.029	0.028	0.025	0.024	0.025
5. KJ461547 Bitylenchus hispaniensis	0.174	0.166	0.008	0.023		0.009	0.000	0.000	0.000	0.008	0.008	0.022	0.025	0.020	0.019	0.020
6. KJ461546 Bitylenchus hispaniensis	0.226	0.214	0.023	0.023	0.036		0.009	0.009	0.010	0.015	0.015	0.030	0.025	0.026	0.026	0.020
7. KJ461545 Bitylenchus hispaniensis	0.173	0.169	0.006	0.022	0.000	0.036		0.000	0.000	0.008	0.008	0.022	0.025	0.020	0.019	0.020
8. KJ461548 Bitylenchus hispaniensis	0.175	0.161	0.008	0.023	0.000	0.038	0.000		0.000	0.008	0.008	0.022	0.025	0.020	0.019	0.021
9. KJ461544 Bitylenchus hispaniensis	0.177	0.176	0.005	0.023	0.000	0.043	0.000	0.000		0.009	0.009	0.024	0.028	0.022	0.021	0.024
10. DQ328707 Bitylenchus dubius	0.185	0.181	0.032	0.034	0.031	0.070	0.030	0.031	0.032		0.000	0.028	0.028	0.022	0.023	0.024
11. EU368590 Bitylenchus dubius	0.185	0.181	0.032	0.034	0.031	0.070	0.030	0.031	0.032	0.000		0.028	0.028	0.022	0.023	0.024
12. KJ461549 Bitylenchus iphilus	0.155	0.171	0.142	0.159	0.124	0.167	0.122	0.125	0.132	0.154	0.154		0.018	0.014	0.013	0.015
13. KJ461526 Paratrophurus striatus	0.173	0.177	0.142	0.151	0.134	0.134	0.132	0.134	0.147	0.154	0.154	0.090		0.013	0.014	0.013
14. KJ461552 Bitylenchus maximus	0.142	0.142	0.117	0.132	0.106	0.142	0.105	0.106	0.117	0.122	0.122	0.072	0.063		0.005	0.010
15. KJ461551 Bitylenchus maximus	0.146	0.151	0.116	0.130	0.103	0.146	0.102	0.103	0.110	0.126	0.126	0.068	0.068	0.017		0.012
16. KJ461533 Bitylenchus brevilineatus	0.160	0.168	0.120	0.135	0.110	0.109	0.108	0.111	0.124	0.128	0.128	0.077	0.061	0.048	0.056	

Fig. (12). Genetic pairwise distance of the 28S rDNA of *Bitylenchus* populations using MEGA11.

Genetic distance measures the genetic divergence between species and populations. Populations with similar alleles or nucleotides have small genetic distances, indicating close relations and a recent common ancestor. A genetic distance of zero means that there is no difference in nucleotides and a complete match. One of the methods for estimating distance is Nei's distance, which is mainly used for comparing populations of the same species. Genetic pairwise distance is a substitution rate of the nucleotides and helps rebuild the history of populations and the origin of biodiversity. When we want to analyze the link between individuals or populations, we will use genetic distance estimation, which calculates the "distance" between samples based on their genetic profile or sequence alignment.

CONSTRUCTING TREES

1. Upload the data file already aligned and saved and choose "analyze".

2. Select "Phylogeny," then "construct/test Neighbour-Joining tree" or "construct/test Maximum Likelihood tree" to display the analysis preferences dialog box.

3. In the options summary tab, click the "test of phylogeny" as "bootstrap" and justify for 2000 or more.

4. Leave the taskbar as software default and press compute to elaborate the tree.

5. After releasing the tree on "Tree Explorer", in "View" menu select "option".

6. Select branch menus and then justify "hide value for less than" for 50.

7. Justify the cut-off value of the condensed tree for ≥ 50.

8. Select the "Image" and save the tree as a jpeg or PDF file.

Many platforms offer various algorithms for phylogenetic tree construction, such as Mr. Bayes (Fig. **13**: Example of Bayesian tree), PhyML, and Mega. It is important to ensure that the packages are compatible before using the software.

Maximum Likelihood Tree

The maximum likelihood method is a statistical approach utilized to identify the unknown parameters of a probability model that define the model and encompass rates, differential transformation costs, and even the tree involved in phylogenetic analysis. An example of a commonly used model is the normal population, which includes two parameters - the mean and variance. In regard to the analysis of phylogenetic trees, Matzke underlined the significance of parameters in 2011.

Advantages

This method of analysis is particularly suited for basic data, including DNA sequences, due to its ability to model stochastic processes statistically. The term "stochastic" refers to a random probability distribution or occurrence that can be statistically analyzed but not precisely predicted. Compared to other techniques, this approach demonstrates lower variance and is less influenced by sampling errors. It is also robust and can outperform alternative methods, such as parsimony or distance methods, even when working with short sequences. This method is well-established statistically and has a clear model of evolution, enabling the assessment of data goodness-of-fit. Additionally, it can assess different tree topologies relative to neighbor joining and uses all sequence information as opposed to distance methods. It also accounts for branch lengths more accurately by combining "multiple hits" to provide more realistic branch lengths and decreasing the region of long branch attraction (LBA). Furthermore, this method captures information from sites that would be uninformative under parsimony, making it a comprehensive analytical tool.

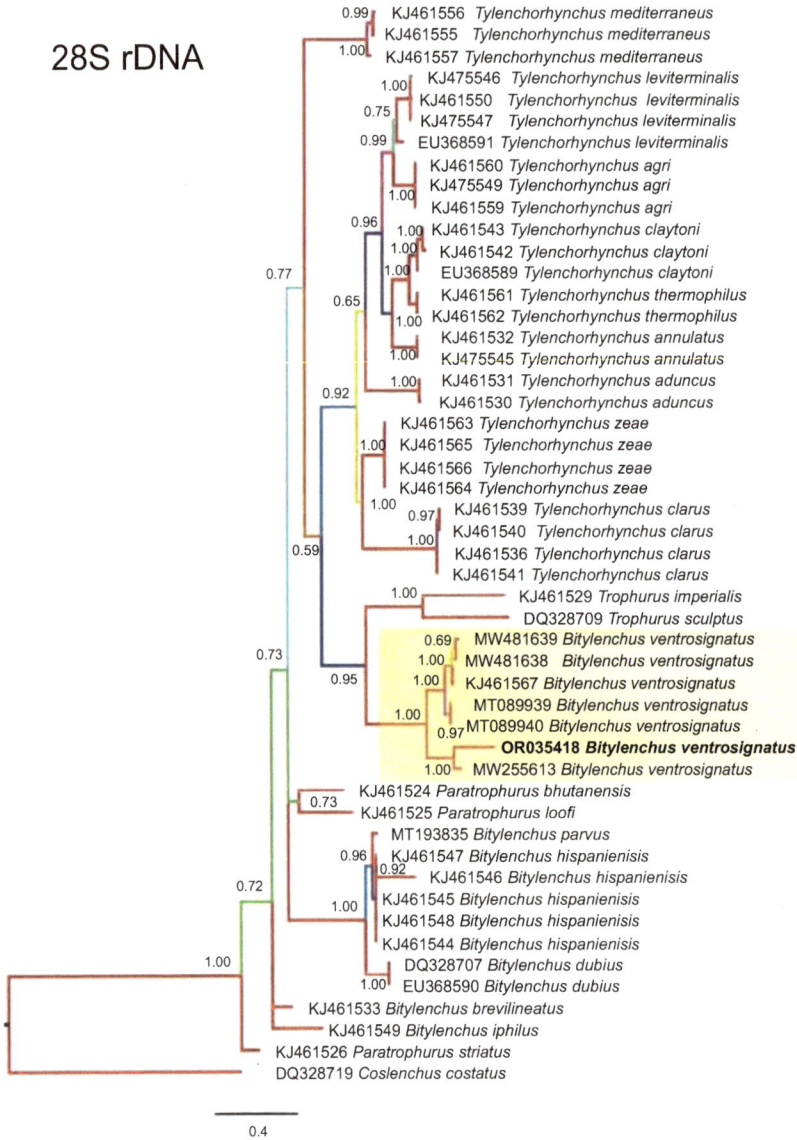

Fig. (13). Phylogenetic tree of 28S rDNA of *Bitylenchus* populations [116].

Disadvantages

1. While the computational process can be intense and result in slower processing times, this is typically a minor issue.

2. It was observed that uneven data distribution in partitions can leave machine learning vulnerable.

3. The model chosen is a crucial factor in determining the outcome, and irrelevant information may be included due to the parsimony principle in model selection.

4. The use of ML to analyze complex data, such as morphology, can be uncertain due to the challenge of modeling numerous processes.

5. The calculation of likelihood equations for a specific distribution can present a significant challenge.

6. For small sample sizes, maximum likelihood estimations may not be reliable.

FEATURES OF A PHYLOGENETIC TREE

External nodes, as indicated in the tree above, include *Bitylenchus* sequences. On the other hand, nematode species sequences are also considered Operational Taxonomic Units or OTUs. The term OTU refers to clusters of uncultivated or unknown microorganisms grouped by DNA sequence similarity of a specific taxonomic marker gene, *e.g.*, rDNA or mtDNA. Internal nodes represent inferred ancestral units considered hypothetical taxonomic units or HTUs. Internal branches that do not end with a tip and external branches that end with a tip occur. External branches are more recent, and internal branches are more ancient. A sister group or sister taxon is a phylogenetic term denoting the closest relatives of another given unit in an evolutionary tree. In the presented tree, *Bitylenchus* and *Trophurus* are considered sister groups. Outer branches representing the taxa or sequences are called leaves in phylogenetic trees.

Finally, in each phylogenetic tree, a group of organisms is believed to comprise all the evolutionary descendants of a common ancestor known as a clade (Fig. **14**).

Phylogenetic trees, also known as Cladograms or Dendrograms, can be categorized as either rooted or unrooted. A rooted tree has a specified root node from which all other nodes descend, establishing a direction corresponding to evolutionary time. This direction allows us to determine ancestor-descendant relationships between nodes. In a rooted tree, when two nodes are connected by a branch, the node closest to the root is considered the ancestor, and the node further away from the root is the descendant. On the other hand, unrooted trees do not have a designated root and do not provide the same information on evolutionary relationships. They do not indicate ancestors or descendants, and sequences adjacent to an unrooted tree may not be closely related evolutionarily. On the other hand, unrooted trees are significant and informative when you want to know a network of relationships between units, including species or populations, without consideration of an evolutionary history (Fig. **15**).

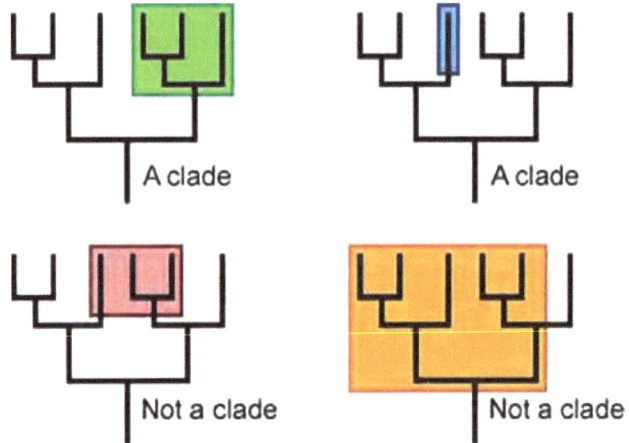

Fig. (14). Schematic view of the clade and non-clade groups in phylogenetic trees.

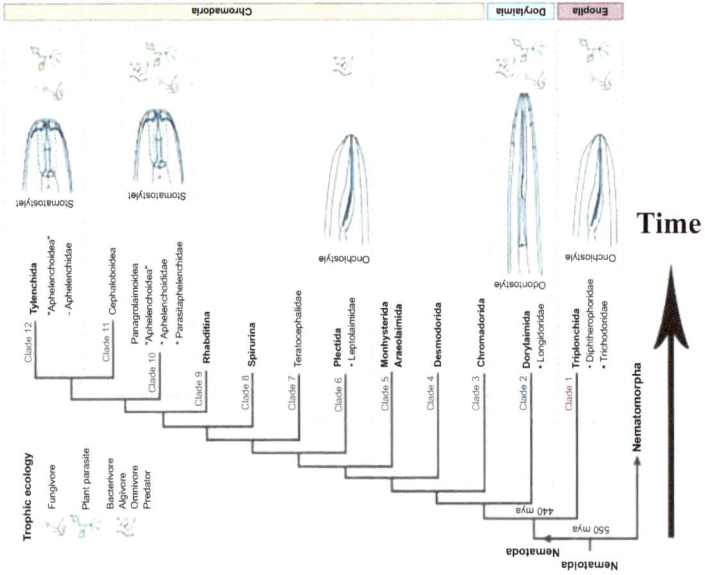

Fig. (15). Schematic overview of the evolution of the trophic behavior of nematoda [117].

BOOTSTRAPPING

(According to MEGA tutorial)

Determining a trustworthy tree can be achieved through Felsenstein's bootstrap test, which evaluates the reliability of the tree using Efron's bootstrap resampling method. Let's say there are "m" sequences, each with "n" nucleotides, and a phylogenetic tree is constructed using a tree-building method. To construct a new tree, "n" nucleotides are randomly selected with replacement from each sequence, generating a new set of sequences. The original tree-building method is then used to reconstruct a new tree using the new nucleotide sequences. The topology of the new tree is compared to that of the original tree, with interior branches in the original tree that differ from the bootstrap tree assigned a score of 0, while all other interior branches receive a score of 1. This process is repeated multiple times, and the percentage of times each interior branch receives a score of 1 is recorded as the bootstrap value. The topology at that branch is considered trustworthy only if the bootstrap value for a particular interior branch is 95% or higher. However, there are more resampling tests for phylogeny, such as jackknifing and cross-validation. In summary, bootstrapping is an approach for estimating the variance. Jackknifing is a method for reducing the bias of an estimator as well as assessing the variance of an estimator. Cross-validation is a method for calculating the error involved in making predictions. On the other hand, bootstrapping is the most helpful and popular method of resampling in phylogenetic studies. The usefulness of Mega in the phylogenetic analysis was explained in this chapter. Still, many other software and techniques, such as Bayesian, PhyML, *etc.*, are used to study the species' phylogenetic analysis and molecular relationship.

OUTGROUP OF PHYLOGENETIC ANALYSIS

In phylogenetic analysis, an out-group is utilized as a reference point to determine the evolutionary relationships among three or more monophyletic groups of organisms. The out-group is carefully chosen to be comparable to the ingroup taxa in a rooted tree. Ideally, the selected out-group is closely related to the other groups but less closely related than any single one of the other groups is to each other. It is important to note that the out-group should not be the same as the ingroup, but rather related to it, in order to demonstrate their relationship.

ONLINE TOOLS FOR NEMATODE IDENTIFICATION

Two online platforms have been developed to address nematode-related issues for diagnosis. The first platform, PPNID, is a molecular pipeline database for identifying plant-parasitic nematodes [118]. The second platform, NemaRec, is an online pipeline for identifying nematodes based on their images [119].

PPNID utilizes the inclusion of the alignment function in the analysis to facilitate the exploration of mutation distribution, unveiling distinct genetic characteristics at the nucleotide level that are essential for discerning between different species. Through the application of genetic distance analysis, graphical plotting, and maximum likelihood phylogeny analysis, a more robust set of criteria can be established to yield optimal results in comparison to similarity searching. This method also contributes to the identification and delineation of species based on evolutionary or phylogeny-based concepts. According to the main source of the publication, the executable file, along with tutorials, is available at https://github.com/xueqing4083/PPNID.

NemaRec utilizes advanced deep convolutional neural networks to analyze nematode images and calculate feeding types, c-p values (for free-living nematodes), Maturity Index, and Plant Parasitic Index for environmental assessment. In the preliminary development for 19 genera, the model accurately identified up to 60% of genera, 76% of c-p values, and 76% of feeding types in the specimen-based dataset from the images captured from microscopy and 94%–97% in the augmented dataset. NemaRec offers high-throughput online identification and collects images uploaded by users for potential model training [119]. According to the main source of publication, the NemaRec is available at http://168.138.167.251:8080.

CONCLUSION

Bioinformatics is an interdisciplinary field that combines biology, computer science, and statistics to analyze and interpret biological data, particularly DNA sequences. Specifically, nematologists use bioinformatics as a tool to analyze the DNA sequences of nematodes and determine their evolutionary relationships through phylogenetic analysis. However, different methods exist for analyzing DNA sequences, and it is crucial to understand how these methods apply to the target group of nematodes to obtain accurate and meaningful results. Therefore, by employing the appropriate bioinformatics methods, nematologists can gain a deeper understanding of nematode evolution and biology.

CHAPTER 8

Biodiversity Analysis

Abstract: This chapter delves into the assessment of nematode biodiversity, which involves investigating the community of these organisms present in a given region. The biodiversity of nematodes can be studied through molecular or conventional approaches. In a conventional method, proper sampling is essential to accurately evaluate biodiversity, and this task is aided by the use of various indices, including Shannon, richness, evenness, and community indices. In the field of molecular biodiversity, researchers can investigate the diversity of nematodes by conducting metagenomic analysis of ribosomal DNA (*e.g.*, 18S rDNA) or cytochrome c oxidase subunit I (cox1) gene of mitochondrial DNA. These approaches provide significant insights into the identification and classification of nematode taxa present in soil ecosystems. Analysis of these genetic markers allows scientists to better understand the rich diversity of nematodes and their ecological roles within soil communities. In this chapter, relevant information on nematode biodiversity assessment is presented.

Keywords: Biodiversity, Free-living, Indices, Nematodes, Plant-parasitic.

CONVENTIONAL BIODIVERSITY

Sampling

The sampling pattern depends on the type of soil (*e.g.*, forest, agricultural, and natural fields). For deep sea, the samples should be taken from different depths [120]. Soil sediments from the deep sea can be carried by using a spade-corer at a specific time. The spade-corer with a 6-cm diameter collects the deposits after 7 hours [120]. For meiofaunal analyses, additional sediment samples can be collected by employing collectors connected to six large cages placed on the seafloor and left for 24 h before recovery [121].

Soil sampling should be done systematically across the area of the study. For instance, collect ten samples from each 40 × 5 m split at approximately 4m intervals. Two parallel zig-zag lines are marked along the field to take the samples, with ten equally spaced sampling points on each line. For each sample, two cores are taken at 30 cm depth (to cover the most free-living and plant-parasitic nematodes) at opposite points on the zig-zag line, using a carbon steel tube or an auger. Then, combine the soil to make up a quantity of about 500 g of soil. Soil samples should be bagged and sealed on a bright plastic carrier to avoid desiccation but keep them out of direct sunlight. Samples can be temporarily stored in a protected box for transportation to the laboratory and then at 4° C until extraction, which should be performed as soon as possible. In our nematology laboratory at the University of Limpopo, about 100-300 gr of soil for counting nematodes and about 200 gr for analytical purposes should be prepared. For seasonal fluctuation, soil samples should be taken from different seasons in the exact location for each season over several years. This is very important when you want to study the seasonal fluctuation of a specific plant-parasitic nematode in a particular locality. Geographical coordinates using the global position system or GPS should be noted for each sampling site.

Nematodes can be extracted from a specific quantity of soil taken from each composite sample using the Baermann funnel technique [122]. After 72 h of incubation at room temperature (25 °C), nematodes can be collected and counted using a stereo microscope/light microscope, and the average populations will be determined. Subsequently, nematodes should be fixed with a hot 4% formaldehyde solution and transferred to anhydrous glycerine [13, 123] method for species identification.

Evaluation of Soil Samples

Rhizosphere soil samples from different fields should be analyzed for pH and electrical conductivity (EC), soil texture, C:N ratio, and various mineral analyses, which can be determined by the standard method [124].

Counting of Nematodes

A counting dish or counting slide Fig. (**1**) is used for nematode counting. When the nematodes have been extracted from the soil and collected in a petri dish, adjust all containers to a specific volume of water, *e.g.,* 10 ml, 20 ml, *etc.* Then, stir the container or tubes using a magnetic stirrer to proportionate the nematodes in the water. Take 1 ml of the solution (or more, depending on the counting slide or counting dish) and count the nematodes. The number of nematodes should be multiplied by the total volume of the solution. For instance, if we have 20 ml of the solution, then the number of individuals in 1 ml should be multiplied by 20. The counting process should be repeated at least five times to obtain an average and make the analysis statistically acceptable. The results can be transferred to an Excel file for further study when the average for all sampling sites is acquired. The schematic view of nematode biodiversity analysis is given in Fig. (**1**).

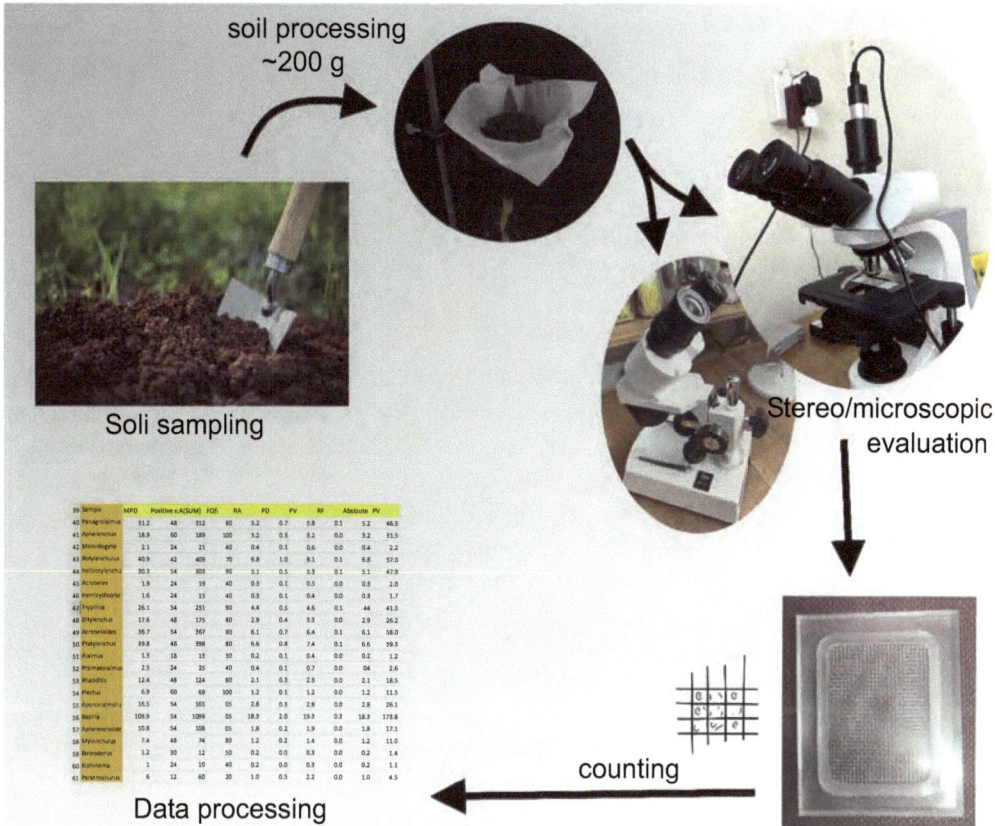

Fig. (1). Schematic view of nematode biodiversity analysis

Biodiversity indices and data analyses

The prominence value (PV) per unit area is commonly used to calculate the correlation between the population density of nematodes (MPD) and the frequency of their occurrence (FO%) based on their genus/species [125]. PVs serve as an index for classifying the nematode genera/species found during surveys [125]. Thus, high PVs for nematode genera indicate the most abundant ones in the rhizosphere soil samples collected from each site. The following equation [126, 127] is used to calculate PV:

PV = Population density × √frequency of occurrence /10

Nematode biodiversity indices represent the Evenness Index (E) [128], Richness Index (SR) [128], Shannon Index (H) [129], and Simpson Index (S) [129].

Shannon index

This index indicates a community's diversity, and it is calculated in the following equation:

$$H' = -\sum p_i \ln p_i$$

In a thoroughly sampled community, we can estimate the proportion of individuals in a species by using the formula $p_i = n_i/N$. Here, p_i represents the proportion of individuals in species i, n_i represents the number of individuals in species i, and N represents the total number of individuals in the community. By taking the inverse of the sum, we can make all summation terms negative because p_i falls between zero and one. In most ecological studies, the values range between 1.5 and 3.5 and rarely exceed 4. The Shannon index increases as the richness and evenness of the community increase. The index's inclusion of both biodiversity components is a strength but can also be seen as a weakness since it makes it challenging to compare communities with significant differences in richness.

Simpson's Index

The Simpson's index measures the probability of two individuals from a large community being of the same species. It is calculated using the following equation: $D = (\sum n (n - 1) / (N (N - 1))$

where N = the total number of organisms of all species and n = the total number of organisms of a particular species.

Practically, with this index, 0 represents unlimited diversity, and 1 represents no diversity. The bigger the value of D, the lower the diversity. This is neither intuitive nor logical, so to get over this problem, D is often subtracted from 1 to obtain Simpson's Index of Diversity, calculated by "$1 - D$".

The index value ranges from 0 to 1, and a higher score indicates greater diversity within the sample. In essence, the index measures the probability of two individuals randomly selected from the sample being from different species.

Richness

The concept of species richness is determined by the number of species detected in a given sample. The higher the number of species detected, the greater the richness of the sample. It is important to note, however, that species richness only takes into account the number of species present and not the abundance of each species. This means that regardless of how many individuals of a particular species are present, all species are attributed the same importance in this measure.

Evenness

The notion of evenness is utilized to assess the equitable distribution of various species in a specific area. This offers an understanding of the proportional abundance of each species concerning the overall diversity of the region.

For example, in Table **1**, we presented three different fields for nematode biodiversity. The number of individuals per sampling site (nr. 1, 2, and 3) is given for each genus. All three samples have the same richness (10 species), and the different total number of individuals varies between 82, 122, and 149 for the sampling sites 1, 2, and 3, respectively. However, the third sample site (nr. 3) has more evenness than the other two. This is because the total number of individuals in the sample is quite evenly distributed between the ten species. Sample 1 is, therefore, considered to be less diverse than samples 2 and 3.

Table 1. Nematode diversity and individual numbers in sampling sites.

-	Sample Number		
Genera	1	2	3
Acrobeles	1	2	3
Acrobeloides	10	12	15
Eucephalobus	15	20	20
Panagrolaimus	20	25	30
Meloidogyne	10	15	15
Pratylenchus	2	2	5
Cruznema	3	6	1
Aphelenchus	10	20	25
Ditylenchus	5	10	15
Prismatolaimus	6	10	20
Total number	82	122	149

Besides, ecological indices Sigma Maturity Index [SMI] and Plant Parasitic index [PPI], functional indices (Enrichment Index [EI], Structure Index [SI], Basal Index [BI] and Channel Index [CI]), and metabolic footprints are calculated and statistically analyzed using the online program NINJA [130]. For example, Shokoohi [131] indicated various indices related to nematode diversity in tomato fields in South Africa (Table **2**).

Table 2. Nematode community indices calculated by NINJA in tomato fields in Limpopo Province, South Africa [131].

Index Name	Field-1	Field-2	Field-3	Field-4
Basal Index	31.5 ± 1.6	26.3 ± 0.5	13.6 ± 1.1	31.7 ± 1.0
Bacterivore footprint	44.7 ± 2.5	77.9 ± 3.2	61.2 ± 5.6	48.0 ± 2.6
Channel Index	60.7 ± 2.9	39.1 ± 1.3	11.5 ± 1.2	86.1 ± 3.4
Composite footprint	385.0 ± 42.4	216.8 ± 9.4	166.5 ± 13.7	176.8 ± 4.2
Enrichment footprint	31.4 ± 5.0	81.3 ± 1.7	48.5 ± 5.1	21.7 ± 1.7
Enrichment Index	46.8 ± 3.1	65.9 ± 0.8	79.4 ± 1.6	33.8 ± 1.3
Fungal/Bacterial	0.9 ± 0.1	1.4 ± 0.0	0.3 ± 0.0	0.7 ± 0.0
Fungivore footprint	25.7 ± 3.4	47.1 ± 1.0	11.9 ± 0.9	22.2 ± 1.8
Herbivore footprint	247.1 ± 41.4	5.2 ± 1.1	25.2 ± 3.8	36.7 ± 3.4
Maturity Index	2.4 ± 0.1	2.1 ± 0.0	2.2 ± 0.1	2.6 ± 0.0
Maturity Index 2-5	2.5 ± 0.0	2.4 ± 0.0	2.8 ± 0.1	2.6 ± 0.0
Omnivore footprint	67.6 ± 6.4	30.6 ± 6.5	68.2 ± 9.5	69.9 ± 2.6
Plant Parasitic Index	2.9 ± 0.0	2.8 ± 0.0	3.0 ± 0.0	3.4 ± 0.0
Shannon Index (H')	1.8 ± 0.0	2.2 ± 0.0	2.2 ± 0.0	2.2 ± 0.0
Sigma Maturity Index	2.5 ± 0.1	2.1 ± 0.0	2.4 ± 0.1	2.8 ± 0.0
Structure footprint	68.8 ± 6.4	90.6 ± 7.5	69.8 ± 9.6	73.7 ± 2.5
Structure Index	56.4 ± 2.4	46.4 ± 2.0	71.1 ± 3.6	62.2 ± 1.5
Total biomass, mg	2.5 ± 0.3	0.8 ± 0.1	0.6 ± 0.1	0.7 ± 0.0
Total number, individual	486.9 ± 31.5	795.4 ± 17.6	568.3 ± 43.2	501.3 ± 14.6

MOLECULAR BIODIVERSITY

Molecular biodiversity is a method used to study nematode communities by using DNA barcodes. Two useful techniques for studying molecular diversity are shotgun metagenomics and high-throughput sequencing. Shotgun metagenomic sequencing allows for a comprehensive understanding of genes from all microorganisms present in a complex sample [132]. So, most of the nematodes will be distinguished, which can give us valuable insight into the diversity of various trophic groups of nematodes in the soil [133].

High-throughput sequencing encompasses a diverse set of techniques designed to analyze large quantities of DNA, RNA, or protein. These methods leverage a range of sequencing mechanisms, including sequencing by synthesis, sequencing by ligation, or the detection of electrical changes. Another term often used interchangeably with high-throughput sequencing is next-generation sequencing (NGS) [134].

NGS is the most popular workflow for the molecular biodiversity of nematodes [135, 136, 137], which includes nucleic acid extraction, library preparation, sequencing, and data analysis and interpretation. The steps are illustrated in Fig. (**2**).

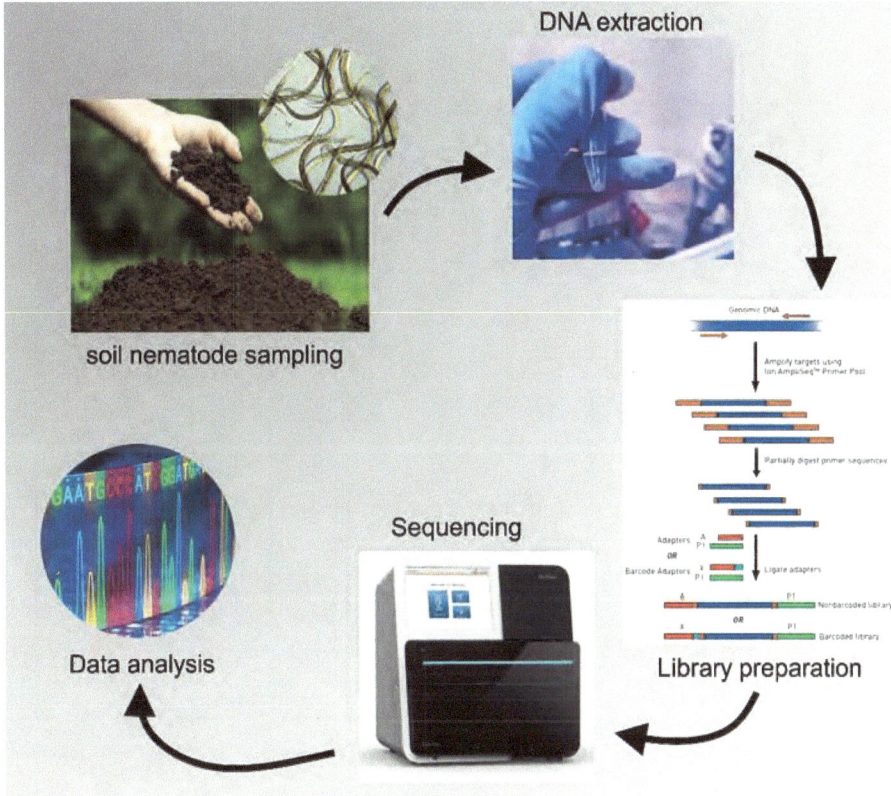

Fig. (2). Next-generation sequencing steps used for molecular biodiversity of nematodes.

Various DNA barcodes have been used for nematode metagenomics analysis. One of the most efficient barcodes is 18S rDNA [135]. The result indicated that among three habitat types, including conventional, organic, and no-tillage cropping systems, families Rhabditidae and Tylenchidae were the most dominant nematodes. The metagenomic study on the soil using 18S rDNA revealed that Nematoda was the third dominant group of eukaryote after Fungi and Arthropoda Fig. (**3**).

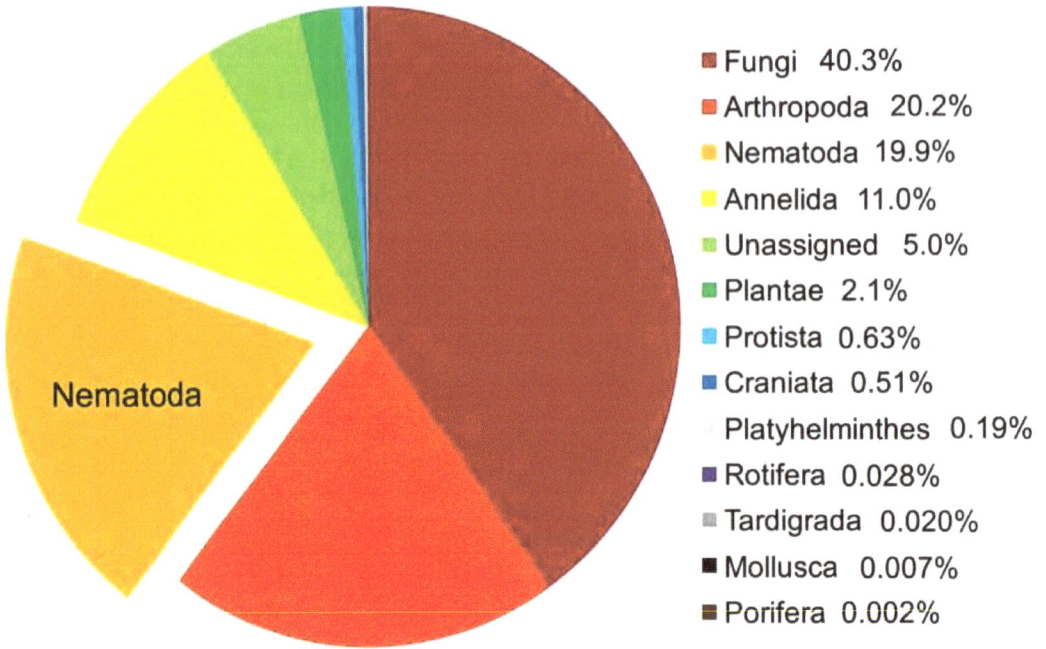

Fig. (3). The eukaryote distribution of the soil using 18S rDNA metagenomic approaches (adopted from Treonis *et al.* [135]).

In a comprehensive study, the nematode community in three distinct soil samples was meticulously analyzed using advanced high-throughput sequencing of 18S rDNA, as illustrated in Fig. (**4**). The detailed results of the analysis unequivocally demonstrated that the group Rhabditida emerged as the most prevalent and dominant among the nematodes present in the soil [136]. The findings further underscored the importance and efficacy of employing metabarcoding as a powerful and indispensable tool for unraveling the intricate molecular diversity of nematodes. It should be noted that four replicates of each location are necessary for NGS analysis of nematode microbiome and biodiversity analysis.

In addition, mtDNA plays a critical role in nematode metabarcoding analysis [137]. New primers for the coxi region of mt DNA, including forward primer (COIFGED: CCTTTGGGCATCCNGARGTNTAT) and reverse primer (JB5GED: ACCTAAACTTARWACRTARTGAAAATG) detected the most nematodes within the soil [137]. By examining the metagenomic analysis of nematodes, it is evident that the 18S rDNA serves as the more favorable barcode, as depicted in Fig. (**5**). This observation implies that there might be a need for more in-depth exploration of mtDNA to ensure the comprehensive detection of all taxa in metagenomic analysis.

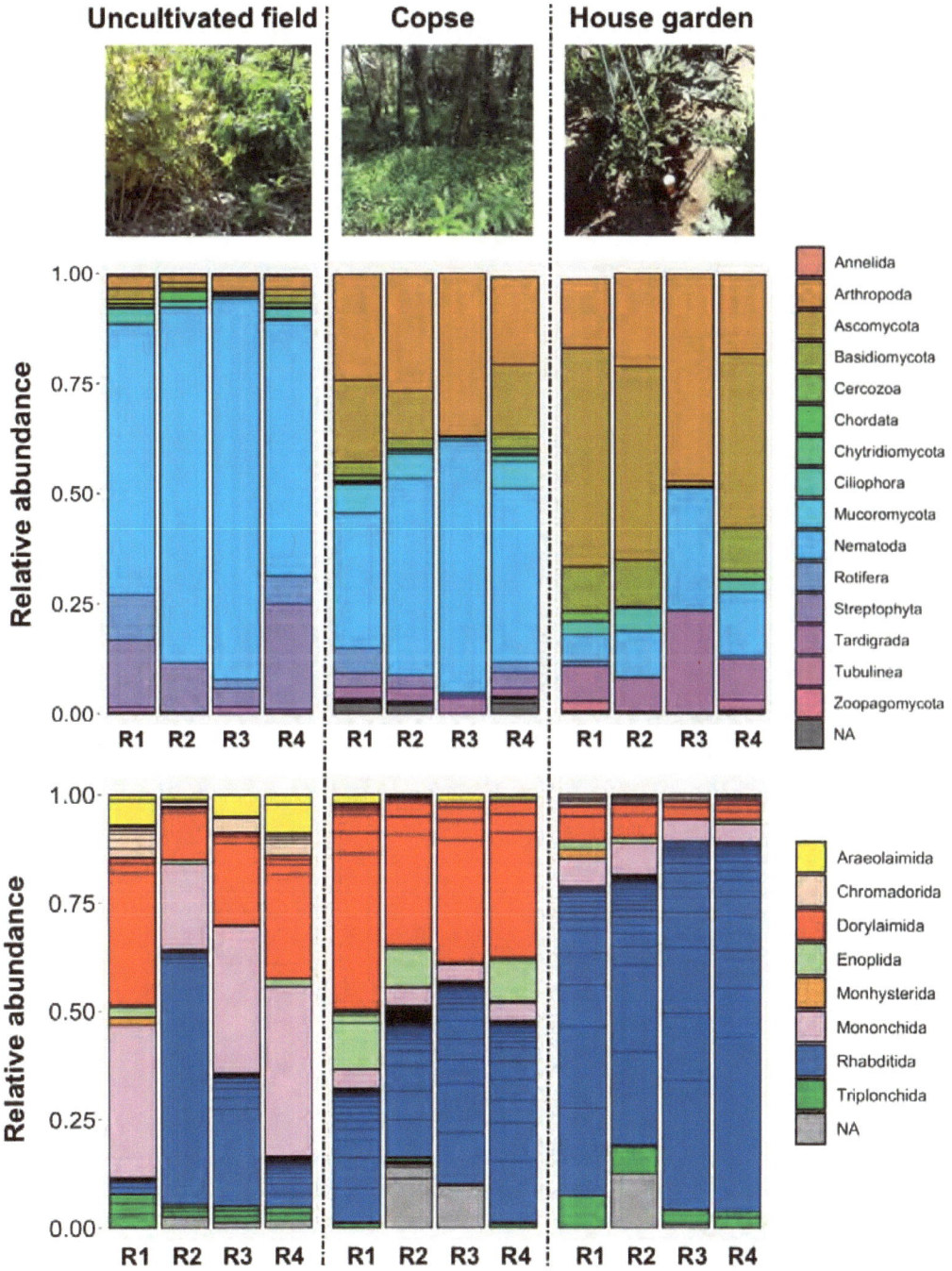

Fig. (4). Diversity of nematodes in different soil types using 18S rDNA metabarcoding (adopted from Kenmotsu *et al.* [136]).

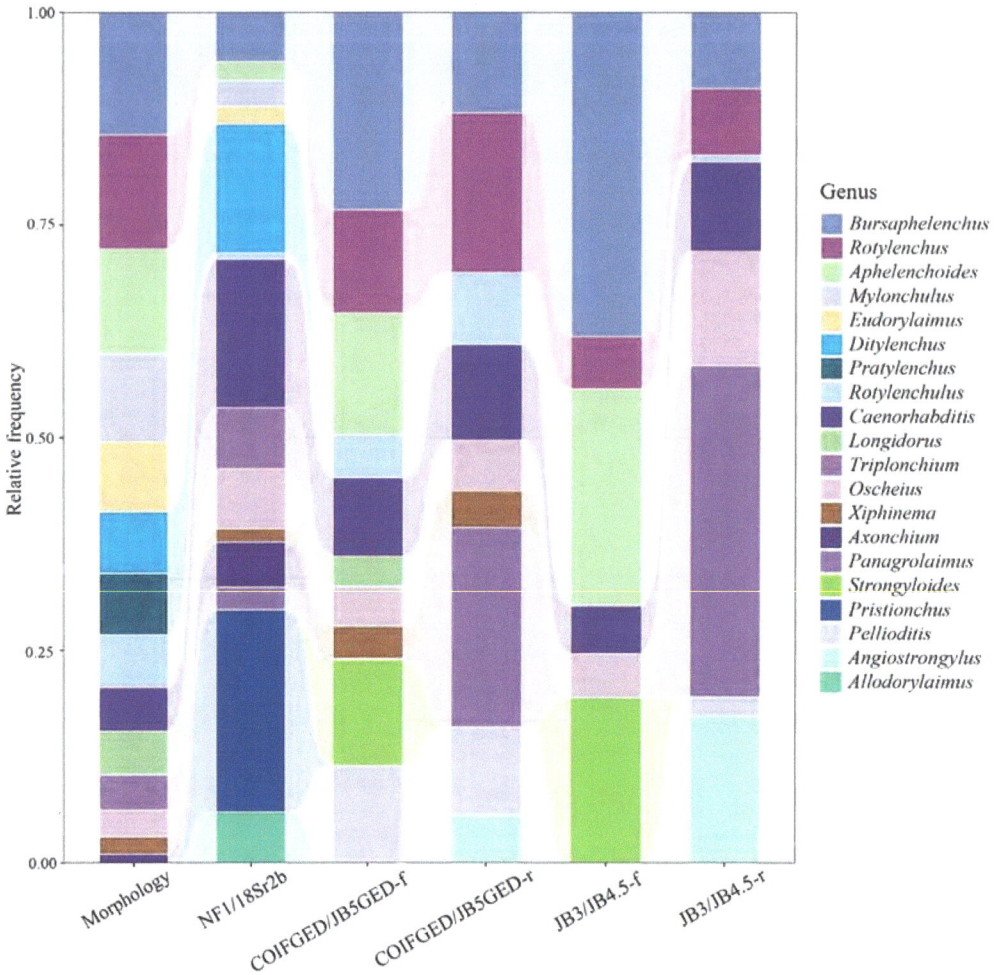

Fig. (5). Nematode community of the soil using different primers and morphology (adopted from Ren *et al.* [137]).

The microbiomes associated with nematodes offer valuable insights into the complex interactions between these organisms and their microbial communities. Over the years, researchers have extensively studied the microbiomes of various nematode species, including *Zeldia punctata* [138], *Acrobeles complexus* [139], *Caenorhabditis elegans* [140], and microhabitat nematodes [141]. This research has provided critical information for understanding the intricate ecological relationships between bacteria and nematodes, particularly in the context of soil biodiversity. The findings from these studies have significantly contributed to our knowledge of microbial ecology and the coexistence dynamics within soil ecosystems.

CONCLUSION

Nematologists have long recognized the importance of biodiversity in understanding the ecology of nematodes. Analyzing the nematode community and its relationship with the environment requires the use of various indices, which makes identifying nematodes the foremost priority in generating initial data for biodiversity analysis. To achieve this, the most efficient way of sampling the area of study must be employed to cover the most nematodes possible. Additionally, understanding the soil's physicochemical properties is paramount in comprehending the relationship between nematodes and their environment. This information is crucial in understanding the impact of climate change and other environmental factors on nematode diversity. By taking these steps, nematologists can glean valuable insights into the intricate relationships between nematodes and their environment, helping to manage the harmful nematodes for future soil health and a sustainable environment.

CHAPTER 9

Nematode Rearing and Greenhouse Studies

Abstract: The maintenance of live nematodes is a crucial task for various studies related to their biology, taxonomy, and genetics. It is essential to have a living source of nematodes for future research purposes. Researchers typically use media or plants to culture and multiply nematodes. While agar serves as a base ingredient for several nematodes, liquid or solid culture media is used for laboratory rearing. It is important to note that culturing plant-parasitic nematodes is more challenging in the lab than free-living nematodes. However, entomopathogenic nematodes can be cultured in large quantities for pest control purposes. This chapter provides a thorough discussion of the culturing methods of several nematodes. Greenhouses are an essential tool for nematologists to study nematodes in the field of integrated pest management. By using various types of greenhouses, researchers can achieve their goals in nematode biology. The Phytotron provides the most precise and controlled environment for biological studies, while certain nematodes, such as *Meloidogyne* and cyst nematodes, can be mass-reared in greenhouses for further molecular or biological surveys. This chapter delves into the important tasks associated with greenhouses for economically significant plant-parasitic nematodes.

Keywords: Bacteria, Biology, Culture, Greenhouse, Laboratory, Multiplication, Phytotron, Water agar.

CULTURING OF NEMATODES

General consideration

Some factors must be considered when determining the type of research, as well as molecular or biological approaches. The most integral is the question concerning the number of nematodes required. In vivo and in vitro cultures help maintain colonies of different nematodes. For large-scale rearing of nematodes, in vitro methods are the most practical. For this purpose, nematode species (free-living or plant-parasitic) and maintenance duration are essential. Concerning the free-living nematodes, for instance, *Panagrolaimus*, they could be reared simply in a WA2% (Water Agar medium) even in the absence of bacteria added to the medium. We tested this medium in our laboratory, and after one week, several individuals for the molecular study were collected (Shokoohi, unpublished data). Some methods for rearing specific nematodes are explained. The nematodes that have been carefully raised are utilized to conduct in-depth research in the areas related to biology, molecular taxonomy, and a wide range of physiological studies. The procedure for several nematodes is as follows.

Globodera spp.

To rear *Globodera*, start by disinfecting potato tubers using a 5% NaOCl solution for 4 minutes and then rinse them with water. Next, dry the potato tubers and leave them at room temperature to allow the eyes to develop and shoots to form. Afterward, soak the cysts in root exudate for one to several days. Prepare a closed container with about 200g of dried, clean sand and 30ml of tap water. Place the germinated potatoes in the container. Crush the cysts to extract the eggs and juveniles and add approximately 1000 eggs and juveniles of *Globodera* spp. to the container with sand. Close the lid and store the containers in the dark at 20°C for about 14 weeks. The cysts can be collected after the roots have died [142].

Heterodera spp.

To prepare for the cyst nematode of *Heterodera*, start by obtaining white mustard seeds (Sinapis alba). Then, sterilize the seed surface by using a 20% dilution of 3.6% sodium hypochlorite for 20 minutes and wash the seeds six times with sterile double-distilled water. Next, keep the seeds at 4°C overnight to enhance and synchronize germination. Prepare the Knop medium by pouring 1 liter of distilled water into a beaker, then adding 3 g of calcium nitrate and 1 g of each of the following chemicals: magnesium sulfate, potassium nitrate, and potassium phosphate, and dissolve them in the water. After that, autoclave the solution. Add 100 g of sucrose to 1 L of Knop's solution, and then soak the seeds in a petri dish containing the Knop medium. For female counting, dye the agar using various food coloring (e.g., Limino, DYL-ghm-20210126–324). For cyst counting, leave the agar undyed. Plants should be grown in a growth chamber with a 16-hour day at 21°C and an 8-hour night at 20°C cycle. Soak the cyst in 3 mM Zinc Chloride to promote egg hatching. Then, inoculate 300 second-stage juvenile individuals to the plants after 21-28 days of plant growth. The cysts will develop in 10–12 weeks at 20–25°C in darkness [143].

Meloidogyne spp.

Root-knot nematodes aiming for molecular diagnosis are cultured in the greenhouse or in an in vivo environment. Therefore, a susceptible cultivar of tomatoes is the best option to rear this nematode. As the molecular diagnosis of root-knot nematodes can be made using eggs, juvenile, and adult stages, therefore, tomatoes should be transplanted into the proper pots and then inoculated with eggs and second-stage juveniles. For obtaining a pure culture of *Meloidogyne*, the egg mass should be isolated from a single female, and then tomatoes must be inoculated. After 56 days, enough nematodes are extracted for various molecular studies [105].

Radophulus spp.

To propagate burrowing nematodes, begin by extracting them from banana roots. Chop the roots into small pieces and collect the nematodes using extraction methods in a beaker. Then, peel the carrots using 70% ethanol and cut them into small 2 cm diameter pieces. Ensure all steps are conducted under laminar flow conditions and use glass Petri dishes to avoid contamination. Collect 60-100 female nematodes in a 10ml measuring cylinder. Add 6 mg of streptomycin sulfate to 10 ml of distilled water to make a sterilization solution for nematodes. Then, in the measuring cylinder, allow the nematodes to settle, and then remove 5 ml of the water. Add 5 ml of the streptomycin solution to the nematode cylinder and let it sit for 1 hour. Repeat this step two more times, leaving the nematodes for 1 hour and then 30 minutes. Reduce the water in the nematode cylinder to 2-3 ml. In a laminar flow hood, add *Radophulus* to the carrot discs. Place 15-50 nematodes per disc using a maximum of three drops of nematode suspension. Seal the Petri dishes with parafilm and incubate for 3-4 weeks at 25-28°C. The presence of callose indicates a successful culture of *Radophulus*. After 3-4 weeks, check the discs under a stereomicroscope without removing the parafilm. If the nematodes are visible, they can be collected [144].

Pratylenchus spp.

Roo-lesion nematodes are very hard to culture on media or in vitro. However, under controlled conditions, these nematodes can be reared in the laboratory. For this purpose, after the isolation of *Pratylenchus* from soil or infected roots, carrot discs are a suitable option for the multiplication of *Pratylenchus* species. It should be noted that all equipment should be sterilized in autoclave. Additionally, carrot discs should also be treated with sterilization materials. Also, for sterilizing nematode, streptomycin sulfate can be used. Next, the nematode should be transferred to the carrot discs, and the Petri dishes should be sealed with parafilm. After 3-4 weeks, *Pratylenchus* can be extracted from carrot discs [144].

Aphelenchoides and *Bursaphelenchus* spp.

To rear foliar nematodes, you should first extract them from the soil, seeds, or leaves. Next, disinfect the *Aphelenchoides* surface with streptomycin sulfate (0.1%) for 10 minutes, followed by rinsing with sterilized water three times. Then, prepare a potato dextrose agar (PDA) and culture some fungi, such as *Alternaria* sp., *Fusarium* sp., and *Botrytis cinerea*. After that, add about 25

Aphelenchoides individuals to each petri dish and incubate them in dark conditions at 20-25°C with the aforementioned fungi. Nematodes are ready for harvest 21 days after infestation [145]. For *Bursaphelenchus xylophilus*, after isolation from pine trees (e.g., *Pinus massoniana*), the nematodes are cultured on PDA plates covered with *B. cinerea* mycelia at 25 °C for 7 days [146].

Mylonchulus spp.

For rearing the predator nematodes, including members of the order Mononchida, an artificial environment, "Soil Extract Agar" (SEA medium), is used. *Mylonchulus* specimens should be extracted from soil, and then a 1.5% SEA must be used for their multiplication. In a Petri dish with SEA medium, five individuals *of M. sigmaturus* adults should be placed. For a better result, about 100 nematodes, such as *Aphelenchus* sp., *Helicotylenchus* sp., and *Ditylenchus* sp. can be used as a food source. Besides, to avoid losing nematodes, it is better to put some water drops around them. It should be noted that all processes should be under controlled conditions to avoid contamination. Then, the Petri dishes are kept in darkness at 25°C. After 50-65 days, a sufficient number of *Mylonchulus* can be isolated [147].

Panagrolaimus spp.

Monoxenic cultures of *Panagrolaimus* sp. can be prepared according to Ayoub *et al.* [148]. The juvenile of *Panagrolaimus* should be transferred to the nematode growth medium (NGM) that already contains *Escherichia coli*. The Petri dishes should be kept at 25 °C for five days. Then, the nematode in Petri dishes can be used in a bigger container to rear the mass of *Panagrolaimus*. In some experiments, after the inoculation of the Petri dishes with adult *Panagrolaimus,* without the presence of *E. coli*, a sufficient number of nematodes were observed.

Caenorhabditis spp.

Caenorhabditis elegans is a model organism, and its culturing in laboratory conditions is critical for various molecular studies [149]. The best medium for in vitro culturing of *Caenorhabditis* is nematode growth medium (NGM), which was used by Brenner [150]. The ingredients for NGM are shown in Table **1** [151].

Table 1. Ingredients of medium for culturing *C. elegans*.

materials	Amount per 1 litre	Final volume
Agar	17.0 g	1.7% (w/v)
NaCl	2.9 g	50 mM
Peptone	2.5 g	0.25% (w/v)

(Table 1) cont.....

materials	Amount per 1 litre	Final volume
CaCl$_2$ (1 M)	1 mL	1 mM
Cholesterol (5 mg/mL)	1 mL	5 µg/mL
KH$_2$PO$_4$ (1 M)	25 mL	25 mM
MgSO$_4$ (1 M)	1 mL	1 mM

* Mix the first three reagents together and autoclave them. After cooling, add the last four reagents to the mixture.

In addition, a mass rearing of *C. elegans* can be done using S-Medium with *E. coli* as a food source [152]. The media and procedure of culturing for different nematodes, which are provided, are examples of being familiar with the methods of multiplication of nematodes. However, many other methods can be used as shown in many scientific works by nematologists/biologists.

GREENHOUSE EXPERIMENTS

The greenhouse is a critical part of the economically important plant-parasitic nematodes such as root-knot and cyst nematodes. Besides, the greenhouse is a controlled situation, which makes it an excellent environment to analyze the interaction of plant-parasitic nematodes and biotic/abiotic factors. In fact, greenhouse and laboratory studies are two steps covering each other to achieve the best result in practical plant nematology. Additionally, the greenhouse is critical for elucidating the relative host suitability and testing the temperature, moisture, nematode development, reproduction, and survival in the realistic environment for nematodes. When evaluating new cultivars, it is important to consider various factors such as their resistance, tolerance, and susceptibility. By doing so, we can determine how well these new cultivars will perform in different conditions and against various threats, such as pests and diseases. This information is crucial for growers and breeders to make informed decisions about which cultivars to invest in and cultivate. On the other hand, greenhouses with accurate environmental controls are recognized as phytotrons and can be used for plants with definite growing conditions or for scientific trials and ecological modeling. In addition, phytotron (Fig. **1**) is used for precise study in plant pathology or nematology, such as gene expression [153]. However, gene expression in tomatoes in the plastic greenhouse might be more than phytotron [154]. Despite the large bench in the greenhouse, a phytotron contains a smaller bench or chamber but is more precise than a regular greenhouse [155].

Fig. (1). A phytotron photograph (Ch Biotech [156]).

Nematode Inoculum

1. Extract *Meloidogyne* (eggs+J2) from the infected roots of the sensitive infected tomato plants

2. Collect the eggs and J2 in tap water in a glass flask, place the flask onto a magnetic stirrer, and keep it in suspension throughout the inoculation process

3. Pipette 10 ml of the suspension into a counting dish

4. Check for at least 95% accuracy level of the conspecific species. If not, repeat the whole process.

5. Adjust the nematode number (1000, 2000, 5000 individuals to be inoculated per seedling/seed in 10 ml water) by diluting it with tap water

Greenhouse Technique for the Evaluation of *Meloidogyne*

1. Before the onset of the experiment, prepare the pots for inoculation of nematodes [105].

2. Fill pots with sterilized soil (a proper soil texture with a 7.5 pH).

3. Transplant four-leaf-stage seedlings of a highly susceptible tomato cultivar (e.g., Moneymaker) [105] before root-knot inoculation into a pot.

4. Inoculate the seedlings of tomatoes with *Meloidogyne* species eggs and J2 by pipetting approximately 1000-2000 of the designated *Meloidogyne* spp.

5. Keep the temperature during the experiment ranging between 19 to 21 °C (min) and 25 to 27 °C (max).

6. Irrigate each pot manually three to four times per week.

7. Terminate the trial 56 days after inoculation (DAI) as *Meloidogyne* spp. can complete at least two generations.

8. Remove the above-ground and clean the root with tap water.

9. Rinse plant roots under tap water to remove debris.

10. Immerse the roots in a 0.1% phloxine-B liquid for 15-20 minutes to dye them and count egg masses.

11. Remove the roots from the staining solution after 20 min, cut them into 1 cm pieces, and transfer them to a beaker with tap water.

12. Check for the red stain inside the roots to count the eggs visible inside the roots.

13. Subsequently calculate the egg-laying females' indices [157]. In order to evaluate the health of the root systems, the study looked for the presence of galls (Fig. **2**) and egg masses. The researchers utilized a rating system, which ranged from 0 to 5. A rating of 0 indicated that there were no galls or egg masses on the roots, while a rating of 1 or 2 meant that 1 or 2 galls or egg masses were present. Ratings of 3 to 5 indicated increasingly severe levels of infestation, with 5 being the most severe, indicating more than 100 galls or egg masses per root system. This rating system provided an effective and precise way to evaluate the condition of the root system, resulting in a more accurate analysis of the data gathered during the study. It should be noted that several indices of root galling have been used for this purpose in the same way presented above.

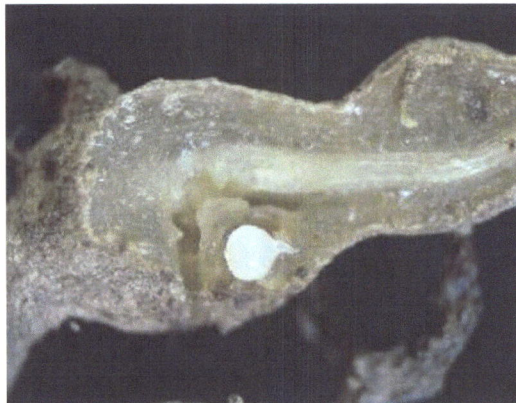

Fig. (2). *Meloidogyne hapla* on the roots of Kiwi, Limpopo Province, South Africa (Photo by: Ebrahim Shokoohi).

Rearing *Meloidogyne* Males

1. Soak tomato seeds in 5-6% sodium hypochlorite for 9-10 minutes, followed by 1-2 minutes in 75% ethanol.

2. To ensure cleanliness, rinse the roots with double distilled water for 4-7.

3. To start your planting, it is recommended to sow the seed in pots. Before planting, make sure to provide proper irrigation and fertilization.

4. After four weeks, inoculate each seedling by placing 3000 J2 around the base of the roots at a depth of 2 cm.

5. Immediately after inoculation, prune lower leaves using a sterile scalpel, leaving only one apical leaf and the shoot apex.

6. After 2 weeks of inoculation, transfer plants to a hydroponic system. Wash the soil and cover the stem with foam plug and plastic to avoid water. Cut a 2-cm hole in the 500-ml container lid for cautious plant insertion until the foam plug fills the hole.

7. Fill the container with 400 ml of 0.5X Hoagland's salt solution.

8. Insert a 2-mm aquarium tube with a pumice aerator at the internal end through a slit in the lid to the bottom of the container.

9. Cover containers with foil to avoid light. Change the nutrient solution every 3 days.

10. Plants thrive at 25°C with artificial light, gentle aeration, and 60% humidity.

11. Collect males every 4 days by filtering the hydroponic tank through a 250-µ--pore sieve stacked on a 43-µm-pore sieve/25-µm-pore sieve.

12. To observe nematodes, collect them from sieves with mesh size of 25-43 µm in the clean Petri dish and observe them under a stereomicroscope [158].

Reproduction Assessment

Extract the eggs and J2 from an infected host plant's roots using the adapted NaOCl technique. Count the eggs and J2 using a stereomicroscope (with proper magnification). The reproduction rate of nematode populations can be calculated using Oostenbrink's reproduction factor (Rf), which is Rf = final number of eggs and J2s (Pf) / initial number of eggs and J2s (Pi) [159].

CONCLUSION

Nematodes play a significant role in various fields, such as agriculture and biological research. Culturing nematodes is a fundamental process in studying the biology of model organisms like *C. elegans* and entomopathogenic nematodes that are used in biological pest control. To achieve the best media for specific nematodes, it is necessary to investigate various ingredients. This process involves identifying the appropriate conditions, such as temperature and humidity, and ensuring the availability of nutrients. Culturing of nematodes in laboratory conditions allows researchers to investigate various aspects of their life, including their growth, development, behavior, and biology. By studying nematodes, we can better understand their ecological role and impact on the environment. This knowledge can be used to develop strategies for the management of nematodes in various fields. In addition, the study of nematode biology is an essential area of research, as these microscopic animals can have a significant impact on the health of plants and animals. To properly understand their behavior and develop effective management strategies, both field and greenhouse studies are necessary. However, when it comes to researching the effects of environmental factors on nematodes, a controlled greenhouse environment is particularly valuable. With a standard greenhouse, nematologists can better investigate the impact of factors such as temperature, humidity, and light on nematode behavior. This is especially important for research on resistance cultivars and biological control, which require a thorough understanding of nematode behavior under different conditions. Hence, a standard greenhouse is a crucial tool for nematologists working in this field.

REFERENCES

[1]   Abd-Elgawad MMM. Optimizing sampling and extraction methods for plant-parasitic and entomopathogenic nematodes. Plants 2021; 10(4): 629.
[http://dx.doi.org/10.3390/plants10040629] [PMID: 33810236]

[2]   Khan MR, Haque Z. Chapter 2 - Methods of assay and detection of nematodes in plant and soil. Editor(s): Khan MR, Quintanilla M. Nematode Diseases of Crops and their Sustainable Management. Academic Press, 2023; 27-54.
[http://dx.doi.org/10.1016/B978-0-323-91226-6.00022-5]

[3]   Nickle WR. Manual of Agricultural Nematology 1991.

[4]   Kleynhans A. Collecting and preserving nematodes: a manual for nematology 1999.

[5]   Santo GS, Nyczepir AP, Johnson DA, O'Bannon JH. Sampling for Nematodes in Soil Available from: https://www.nwpotatoresearch.com/images/documents/SamplingForNematodesInSoil.pdf

[6]   Kergunteuil A, Campos-Herrera R, Sánchez-Moreno S, Vittoz P, Rasmann S. The abundance, diversity, and metabolic footprint of soil nematodes is highest in high elevation Alpine grasslands. Front Ecol Evol 2016; 4: 84.
[http://dx.doi.org/10.3389/fevo.2016.00084]

[7]   Shokoohi E, Abolafia J, Swart A, Moyo N, Eisenback J. *Mesorhabditis sudafricana* n. sp. (Rhabditida, Mesorhabditidae), a new species with a short tail from South Africa. Nematology 2023; 25(7): 775-90.
[http://dx.doi.org/10.1163/15685411-bja10254]

[8]   Freitas VM, Cares JE, Andrade EP, Huang SP. Influence of *Citrus* spp. on the community of soil nematodes in the dry and rainy seasons in Distrito Federal of Brazil. Nematologia Brasileira Piracicaba (SP) Brasil 2008; 32(1): 20-32.

[9]   Borgonie G, García-Moyano A, Litthauer D, *et al.* Nematoda from the terrestrial deep subsurface of South Africa. Nature 2011; 474(7349): 79-82.
[http://dx.doi.org/10.1038/nature09974] [PMID: 21637257]

[10]   Shokoohi E. Identification of the order Rhabditida from Tehran Province, Iran. Ph.D. thesis. University of Tehran, 2008; 276 pages.

[11]   Chitamber JJ. Protocol for collecting and handling plant and soil samples for the detection of plant parasitic nematodes at the nematology laboratory, plant pest diagnostic branch 2003.

[12]   Elmiligy IA, De Grisse A. Effect of extraction technique and adding fixative to soil before storing on recovery of plant-parasitic nematodes. Nematologica 1970; 16(3): 353-8.
[http://dx.doi.org/10.1163/187529270X00036]

[13]   De Grisse AT. Contribution to the Morphology and the Systematics of the Criconematidae (Taylor, 1936) Thorne, 1949 1969.

[14]   Freckman DW, Kaplan DT, Van Gundy SD. A comparison of techniques for extraction and study of anhydrobiotic nematodes from dry soils. J Nematol 1977; 9(2): 176-81.
[PMID: 19305588]

[15]   Young TW. An incubation method for collecting migratory endoparasitic nematodes. Plant Dis Rep 1954; 38: 794-5.

[16]   Bélair G, Simard L, Eisenback JD. First report of the barley root-knot nematode *Meloidogyne naasi* infecting annual bluegrass on a golf course in Quebec, Canada. Plant Dis 2006; 90(8): 1109.
[http://dx.doi.org/10.1094/PD-90-1109A] [PMID: 30781317]

[17]   Hallman, J. and Viaene, N., PM 7/119 (1) Nematode extraction. EPPO Bulletin, 2013; 43(3): pp. 471-495.

[http://dx.doi.org/10.1111/epp.12077]

[18] Baunacke W. Untersuchungen zur Biologie und Bekämpfung der Rübennematoden, *Heterodera schachtii* Schmidt. Arb Biol Reichsanst Berlin 1922; 11: 185-288.

[19] Coolen WA, Hendrickx G, d'Herde CJ. Method for the quantitative extraction of nematodes from plant tissue and its application in the testing of rose rootstocks for resistance to endoparasitic root nematodes. Rijksstation voor Nematologie en Entomologie, Publicatie No. W6, Wetteren, Belgium 1971; pp 34.

[20] Seinhorst JW. The estimation of densities of nematode populations in soil and plants. Växtskyddsrapporter 1988; No 51, Uppsala (SE).

[21] Shokoohi E. Observation on Rhabditina and Tylenchina nematodes in healthy and rotten woods, and soil in Limpopo Province, South Africa. Unpublished report 2024.

[22] Bedding RA, Akhurst RJ. A simple technique for the detection o f insect paristic rhabditid nematodes in soil. Nematologica 1975; 21(1): 109-10.
[http://dx.doi.org/10.1163/187529275X00419]

[23] White GF. A method for obtaining infective nematode larvae from cultures. Science 1927; 66(1709): 302-3.
[http://dx.doi.org/10.1126/science.66.1709.302.b] [PMID: 17749713]

[24] Orozco RA, Lee MM, Stock SP. Soil sampling and isolation of entomopathogenic nematodes (Steinernematidae, Heterorhabditidae). J Vis Exp 2014; 11(89): 52083.
[http://dx.doi.org/10.3791/52083-v] [PMID: 25046023]

[25] Seddiqi E, Shokoohi E, Divsalar N, Abolafia J. Descriptions of four known species of the families Panagrolaimidae and Alloionematidae (Nematoda: Rhabditida) from Iran. Trop Zool 2016; 29(2): 87-110.
[http://dx.doi.org/10.1080/03946975.2016.1177384]

[26] Esser RP. *Romanomermis culicivorax* Ross and Smith 1976, a nematode parasite of mosquitoes Nematology Circular, Division of Plant Industry. Florida Department of Agriculture and Consumer Services 1980; 65: 2.

[27] Kobylinski KC, Sylla M, Black W IV, Foy BD. Mermithid nematodes found in adult *Anopheles* from southeastern Senegal. Parasit Vectors 2012; 5(1): 131.
[http://dx.doi.org/10.1186/1756-3305-5-131] [PMID: 22741946]

[28] Abagli AZ, Alavo TBC, Perez-Pacheco R, Platzer EG. Efficacy of the mermithid nematode, *Romanomermis iyengari*, for the biocontrol of *Anopheles gambiae*, the major malaria vector in sub-Saharan Africa. Parasit Vectors 2019; 12(1): 253.
[http://dx.doi.org/10.1186/s13071-019-3508-6] [PMID: 31118105]

[29] Elbrense H, Shamseldean M, Meshrif W, Seif A. The parasitic impact of Romanomermis iyengari Welch (Nematoda: Mermithidae) on the survival and biology of the common mosquito, Culex pipiens L. (Diptera: Culicidae). Afr Entomol 2022; 30: 30.
[http://dx.doi.org/10.17159/2254-8854/2022/a11687]

[30] McSorley R. Extraction of nematodes and sampling methods.Principles and Practices of Nematode Control in Crops 1987; 13-47.

[31] Van Bezooijen J. Methods and techniques for Nematology 2006.

[32] Shokoohi E. Microbiome and nematode biodiversity of soil nematodes in riverside of Magoebaskloof, Limpopo Province, South Africa. Unpublished report 2024.

[33] Shokoohi E, Abolafia J. Soil and freshwater Rhabditid nematodes (Nematoda, Rhabditida) from Iran: a compendium 2019.

[34] Chałańska A, Bogumił A, Malewski T, Kowalewska K. The effect of two fixation methods (TAF and DESS) on morphometric parameters of *Aphelenchoides ritzemabosi*. Zootaxa 2016; 4083(2): 297-300.

[http://dx.doi.org/10.11646/zootaxa.4083.2.9] [PMID: 27394233]

[35]   De Ley P. A resource for nematode phylogeny Available from: http://xyala. cap.ed.ac.uk/research/nematodes/fgn/worm/extrafix.html

[36]   Scientifica. A guide to Differential Interference Contrast (DIC). Available from: https://www.scientifica.uk.com/learning-zone/differential-interference-contrast

[37]   de Man JG. Die einheimischen, frei in der reinen Erde und im süssen Wasser lebenden Nematoden. Tijdschrift der Nederlandsche Dierkundige Vereeiniging 1880; 5: 1-104.

[38]   Loof PAA, Coomans A. On the development and location of the oesophageal gland nuclei in the Dorylaimina. Proc IX Int Nem Symp Warsaw. 1970; pp. 1967; 1970; 79-161.

[39]   Jairajpuri MS, Ahmad W. Dorylaimida Free-living, predaceous and plant-parasitic nematodes 1992. [http://dx.doi.org/10.1163/9789004630475]

[40]   Abolafia J. A low-cost technique to manufacture a container to process meiofauna for scanning electron microscopy. Microsc Res Tech 2015; 78(9): 771-6. [http://dx.doi.org/10.1002/jemt.22538] [PMID: 26178782]

[41]   Yoder M, De Ley IT, Wm King I, *et al.* DESS: a versatile solution for preserving morphology and extractable DNA of nematodes. Nematology 2006; 8(3): 367-76. [http://dx.doi.org/10.1163/156854106778493448]

[42]   Shokoohi E, Eisenback J. Description of *Anaplectus deconincki* n. sp. from South Africa. J Helminthol 2023; 97e52 [http://dx.doi.org/10.1017/S0022149X23000330] [PMID: 37395051]

[43]   Nadler SA. DNA extraction protocols 2024. Available from: https://nadlerlab.faculty.ucdavis.edu/lab-protocols-and-databases/dna-extraction-protocols

[44]   Floyd R, Abebe E, Papert A, Blaxter M. The NaOH single-nematode DNA extraction method. Mol Ecol 2002; 11: 839-50. [http://dx.doi.org/10.1046/j.1365-294X.2002.01485.x] [PMID: 11972769]

[45]   Herrmann M, Mayer WE, Sommer RJ. Nematodes of the genus *Pristionchus* are closely associated with scarab beetles and the Colorado potato beetle in Western Europe. Zoology 2006; 109(2): 96-108. [http://dx.doi.org/10.1016/j.zool.2006.03.001] [PMID: 16616467]

[46]   Holterman M, van der Wurff A, van den Elsen S, *et al.* Phylum-wide analysis of SSU rDNA reveals deep phylogenetic relationships among nematodes and accelerated evolution toward crown Clades. Mol Biol Evol 2006; 23(9): 1792-800. [http://dx.doi.org/10.1093/molbev/msl044] [PMID: 16790472]

[47]   Rubtsova TV, Moens M, Subbotin SA. PCR amplification of a rRNA gene fragment from formalin-fixed and glycerine-embedded nematodes from permanent slides. Russ J Nematol 2005; 13(2): 137-40.

[48]   Blok VC, Powers TO. Biochemical and molecular identification. Root-knot nematodes, 2009; 98-118. [http://dx.doi.org/10.1079/9781845934927.0098]

[49]   Wu Y, Peng H, Liu S, *et al.* Investigation and identification of cyst nematodes in the Bashang region of Hebei, China. Agronomy (Basel) 2022; 12(9): 2227. [http://dx.doi.org/10.3390/agronomy12092227]

[50]   Correa VR, dos Santos MFA, Almeida MRA, Peixoto JR, Castagnone-Sereno P, Carneiro RMDG. Species-specific DNA markers for identification of two root-knot nematodes of coffee: Meloidogyne arabicida and M. izalcoensis. Eur J Plant Pathol 2013; 137(2): 305-13. [http://dx.doi.org/10.1007/s10658-013-0242-3]

[51]   Zijlstra C, Donkers-Venne DTHM, Fargette M. Identification of *Meloidogyne incognita, M. javanica* and *M. arenaria* using sequence characterised amplified region (SCAR) based PCR assays. Nematology 2000; 2(8): 847-53. [http://dx.doi.org/10.1163/156854100750112798]

[52]   Randig O, Carneiro RMDG, Castagnone-Sereno P. Identificação das principais especies de *Meloidogyne parasitas* do cafeeiro no Brasil com marcadores SCAR-CAFÉ em Multiplex-PCR. Nematol Bras 2004; 28: 1-10.

[53]   Wishart J, Phillips MS, Blok VC. Ribosomal intergenic spacer: a polymerase chain reaction diagnostic for *Meloidogyne chitwoodi, M. fallax*, and *M. hapla*. Phytopathology 2002; 92(8): 884-92.
[http://dx.doi.org/10.1094/PHYTO.2002.92.8.884] [PMID: 18942968]

[54]   Rusinque L, Nóbrega F, Serra C, Inácio ML. The northern root-knot nematode *Meloidogyne hapla*: New host records in Portugal. Biology (Basel) 2022; 11(11): 1567.
[http://dx.doi.org/10.3390/biology11111567] [PMID: 36358268]

[55]   Meng QP, Long H, Xu JH. PCR assays for rapid and sensitive identification of three major root-knot nematodes, *Meloidogyne incognita, M. javanica* and *M. arenaria*. Zhi Wu Bing Li Xue Bao 2004; 34: 204-10.

[56]   Maleita C, Cardoso JMS, Rusinque L, Esteves I, Abrantes I. Species-specific molecular detection of the root knot nematode *Meloidogyne luci*. Biology (Basel) 2021; 10(8): 775.
[http://dx.doi.org/10.3390/biology10080775] [PMID: 34440007]

[57]   Fullaondo A, Barrena E, Viribay M, Barrena I, Salazar A, Ritter E. Identification of potato cyst nematode species *Globodera rostochiensis* and *G. pallida* by PCR using specific primer combinations. Nematology 1999; 1(2): 157-63.
[http://dx.doi.org/10.1163/156854199508126]

[58]   Qi XL, Peng DL, Peng H, Long HB, Huang WK, He WT. Rapid molecular diagnosis based on SCAR marker system for cereal cyst nematode. Zhongguo Nong Ye Ke Xue 2012; 45: 4388-95.

[59]   Peng H, Qi X, Peng D, *et al.* Sensitive and direct detection of *Heterodera filipjevi* in soil and wheat roots by species-specific SCAR-PCR Assays. Plant Dis 2013; 97(10): 1288-94.
[http://dx.doi.org/10.1094/PDIS-02-13-0132-RE] [PMID: 30722143]

[60]   Ou S, Peng D, Liu X, Li Y, Moens M. Identification of *Heterodera glycines* using PCR with sequence characterised amplified region (SCAR) primers. Nematology 2008; 10(3): 397-403.
[http://dx.doi.org/10.1163/156854108783900212]

[61]   Jiang C, Zhang Y, Yao K, *et al.* Development of a species-specific SCAR-PCR assay for direct detection of sugar beet cyst nematode (*Heterodera schachtii*) from infected roots and soil samples. Life (Basel) 2021; 11(12): 1358.
[http://dx.doi.org/10.3390/life11121358] [PMID: 34947889]

[62]   Hoyer U, Burgermeister W, Braasch H. Identification of *Bursaphelenchus* species (Nematoda, Aphelenchoididae) on the basis of amplified ribosomal DNA (ITS-RFLP). Nachr Dtsch Pflanzenschutzd 1998; 50: 273-7.

[63]   Nega A. Review on Nematode Molecular Diagnostics: From Bands to Barcodes. J Biol Agric Healthc 2014; 4(27): 129-53.

[64]   Harris TH, Szalanski AL, Powers TO. Molecular identification of nematodes manual 2001.

[65]   Ibrahim SK, Perry RN, Burrows PR, Hooper DJ. Differentiation of species and population *Aphelenchoides* and of *Ditylenchus angustus* using a fragment of ribosomal DNA. J Nematol 1994; 26(4): 412-21.
[http://dx.doi.org/10.1163/003525994X00292] [PMID: 19279910]

[66]   Szalanski AL, Sui DD, Harris TS, Powers TO. Identification of cyst nematodes of agronomic and regulatory concern with PCR-RFLP of ITS1. J Nematol 1997; 29(3): 255-67.
[PMID: 19274157]

[67]   Subbotin S, Waeyenberge L, Moens M. Identification of cyst forming nematodes of the genus Heterodera (Nematoda: Heteroderidae) based on the ribosomal DNA-RFLP. Nematology 2000; 2(2): 153-64.

[http://dx.doi.org/10.1163/156854100509042]

[68]   Wendt KR, Vrain TC, Webster JM. Separation of three species of *Ditylenchus* and some host races of *D. dipsaci* by restriction fragment length polymorphism. J Nematol 1993; 25(4): 555-63.
[PMID: 19279809]

[69]   Saiki RK, Gelfand DH, Stoffel S, *et al.* Primer-directed enzymatic amplification of DNA with a thermostable DNA polymerase. Science 1988; 239(4839): 487-91.
[http://dx.doi.org/10.1126/science.2448875] [PMID: 2448875]

[70]   Reid A, Manzanille-López RH, Hunt DJ. *Nacobbus aberrans* (Thorne, 1935) Thorne & Allen, 1944 (Nematoda: Pratylenchidae); a nascent species complex revealed by RFLP analysis and sequencing of the ITS-rDNA region. Nematology 2003; 5(3): 441-51.
[http://dx.doi.org/10.1163/156854103769224421]

[71]   Orui Y. Discrimination of the main *Pratylenchus* species (Nematoda: Pratylenchidae) in Japan by PCR-RFLP analysis. Appl Entomol Zool 1996; 31(4): 505-14.
[http://dx.doi.org/10.1303/aez.31.505]

[72]   Waeyenberge L, Ryss A, Moens M, Pinochet J, Vrain T. Molecular characterisation of 18 Pratylenchus species using rDNA Restriction Fragment Length Polymorphism. Nematology 2000; 2(2): 135-42.
[http://dx.doi.org/10.1163/156854100509024]

[73]   Fallas GA, Hahn ML, Fargette M, Burrows PR, Sarah JL. Molecular and biochemical diversity among isolates of *Radopholus* spp. from different areas of the world. J Nematol 1996; 28(4): 422-30.
[PMID: 19277160]

[74]   Schmitz VB, Burgermeister W, Braasch H. Molecular genetic classification of central European *Meloidogyne chitwoodi* and *M. fallax* populations. Nachr Dtsch Pflanzenschutzd 1998; 50: 310-7.

[75]   Širca S, Stare BG, Strajnar P, Urek G. PCR-RFLP diagnostic method for identifying *Globodera* species in Slovenia. Phytopathol Mediterr 2010; 49: 361-9.

[76]   Vrain TC, Wakarchuk DA, Levesque AC, Hamilton RI. Intraspecific rDNA restriction fragment length polymorphism in the *Xiphinema americanum* group. Fundam Appl Nematol 1992; 15: 563-73.

[77]   Dağlı D, Duman N, Yüksel E, *et al.* Characterization of cereal cyst nematodes in wheat using morphometrics, SCAR markers, RFLP, and rDNA-ITS sequence analyses. Trop Plant Pathol 2023; 48(2): 207-16.
[http://dx.doi.org/10.1007/s40858-022-00528-7]

[78]   Song W, Dai M, Shi Q, Liang C, Duan F, Zhao H. Diagnosis and characterization of *Ditylenchus destructor* isolated from *Mazus japonicus* in China. Life (Basel) 2023; 13(8): 1758.
[http://dx.doi.org/10.3390/life13081758] [PMID: 37629615]

[79]   Kumari S, Subbotin SA. Molecular characterization and diagnostics of stubby root and virus vector nematodes of the family Trichodoridae (Nematoda: Triplonchida) using ribosomal RNA genes. Plant Pathol 2012; 61(6): 1021-31.
[http://dx.doi.org/10.1111/j.1365-3059.2012.02598.x]

[80]   Sigma-Aldrich. Restriction enzymes. Available from: https://www.sigmaaldrich.com/ZA/en/technicaldocuments/technical-article/genomics/sequencing/restriction-enzymes

[81]   Zabeau M, Vos P. Selective restriction fragment amplification: A general method for DNA fingerprinting. European patent Application no. 92402627.7. Publication number EP 1993; 0534858, A:1.

[82]   Vos P, Hogers R, Bleeker M, *et al.* AFLP: a new technique for DNA fingerprinting. Nucleic Acids Res 1995; 23(21): 4407-14.
[http://dx.doi.org/10.1093/nar/23.21.4407] [PMID: 7501463]

[83]   Heun M, Schäfer-Pregl R, Klawan D, *et al.* Site of einkorn wheat domestication identified by DNA

fingerprinting. Science 1997; 278(5341): 1312-4.
[http://dx.doi.org/10.1126/science.278.5341.1312]

[84] Semblat JP, Wajnberg E, Dalmasso A, Abad P, Castagnone-Sereno P. High-resolution DNA fingerprinting of parthenogenetic root-knot nematodes using AFLP analysis. Mol Ecol 1998; 7(1): 119-25.
[http://dx.doi.org/10.1046/j.1365-294x.1998.00326.x] [PMID: 9465419]

[85] Devran Z, Firat AF, Tör M, Mutlu N, Elekçioğlu IH. AFLP and SRAP markers linked to the mj gene for root-knot nematode resistance in cucumber. Sci Agric 2011; 68(1): 115-9.
[http://dx.doi.org/10.1590/S0103-90162011000100017]

[86] Rao U, Rao S, Rathi A, Gothalwal R, Atkinson H. A comparison of the variation in Indian populations of pigeonpea cyst nematode, *Heterodera cajani* revealed by morphometric and AFLP analysis. ZooKeys 2011; 135(135): 1-19.
[http://dx.doi.org/10.3897/zookeys.135.1344] [PMID: 22259298]

[87] De Gruijter JM, Gasser RB, Polderman AM, Asigri V, Dijkshoorn L. High resolution DNA fingerprinting by AFLP to study the genetic variation among *Oesophagostomum bifurcum* (Nematoda) from human and non-human primates from Ghana. Parasitology 2005; 130(2): 229-37.
[http://dx.doi.org/10.1017/S0031182004006249] [PMID: 15727072]

[88] Thermo Fisher Scientific Inc. PCR Setup—Six Critical Components to Consider. 2024; Available from: https://www.thermofisher.com/

[89] Shokoohi E. First report of *Tripylina zhejiangensis* associated with grassland in South Africa. Helminthologia 2022; 59(3): 311-6.
[http://dx.doi.org/10.2478/helm-2022-0025] [PMID: 36694827]

[90] Aliramaji F, Taheri A, Shokoohi E. Description of *Paramylonchulus iranicus* sp. n. (Nematoda: Mononchida) from Iran. J Helminthol 2023; 97e53
[http://dx.doi.org/10.1017/S0022149X23000366] [PMID: 37395177]

[91] Nadler SA, D'Amelio S, Dailey MD, Paggi L, Siu S, Sakanari JA. Molecular phylogenetics and diagnosis of *Anisakis, Pseudoterranova*, and *Contracaecum* from northern Pacific marine mammals. J Parasitol 2005; 91(6): 1413-29.
[http://dx.doi.org/10.1645/GE-522R.1] [PMID: 16539026]

[92] Subbotin SA, Ragsdale EJ, Mullens T, Roberts PA, Mundo-Ocampo M, Baldwin JG. A phylogenetic framework for root lesion nematodes of the genus *Pratylenchus* (Nematoda): Evidence from 18S and D2–D3 expansion segments of 28S ribosomal RNA genes and morphological characters. Mol Phylogenet Evol 2008; 48(2): 491-505.
[http://dx.doi.org/10.1016/j.ympev.2008.04.028] [PMID: 18514550]

[93] Shokoohi E, Mehrabi-Nasab A, Mirzaei M, Peneva V. Study of mononchids from Iran, with description of Mylonchulus kermaniensis sp. n. (Nematoda: Mononchida). Zootaxa 2013; 3599(6): 519-34.
[http://dx.doi.org/10.11646/zootaxa.3599.6.2] [PMID: 24614027]

[94] Shokoohi E, Mehrabi Nasab A, Abolafia J. Study of the genus *Ironus* Bastian, 1865 (Enoplida: Ironidae) from Iran with a note on phylogenetic position of the genus. Nematology 2013; 15(7): 835-49.
[http://dx.doi.org/10.1163/15685411-00002722]

[95] Shokoohi E, Abolafia J, Mehrabi Nasab A, Peña-Santiago R. On the identity of *Labronema vulvapapillatum* (Meyl, 1954) Loof & Grootaert, 1981 (Dorylaimida, Qudsianematidae). Russ J Nematol 2013; 21(1): 31-40.

[96] Zograf JK, Semenchenko AA, Mordukhovich VV. New deep-sea species of *Aborjinia* (Nematoda, Leptosomatidae) from the North-Western Pacific: an integrative taxonomy and phylogeny. ZooKeys 2024; 1189(1189): 231-56.
[http://dx.doi.org/10.3897/zookeys.1189.111825] [PMID: 38282715]

[97] Shokoohi E, Mehdizadeh S, Amirzadi N, Abolafia J. Four new geographical records of rhabditid nematodes (Nematoda: Rhabditida: Rhabditomorpha) from Iran with a note on phylogenetic position of the genus *Pelodera* Schneider, 1866. Russ J Nematol 2014; 22(1): 49-66.

[98] Koohkan M, Shokoohi E, Abolafia J. Study of some mononchids (Mononchida) from Iran with a compendium of the genus *Anatonchus*. Trop Zool 2014; 27(3): 88-127.
[http://dx.doi.org/10.1080/03946975.2014.966457]

[99] Koohkan M, Shokoohi E, Mullin P. Phylogenetic relationships of three families of the suborder Mononchina Kirjanova & Krall, 1969 inferred from 18S rDNA. Nematology 2015; 17(9): 1113-25.
[http://dx.doi.org/10.1163/15685411-00002928]

[100] Shokoohi E, Moyo N. Molecular character of *Mylonchulus hawaiiensis* and morphometric differentiation of six *Mylonchulus* (Nematoda; Order: Mononchida; Family: Mylonchulidae) species using multivariate analysis. Microbiol Res (Pavia) 2022; 13(3): 655-66.
[http://dx.doi.org/10.3390/microbiolres13030047]

[101] He Y, Subbotin SA, Rubtsova TV, Lamberti F, Brown DJF, Moens M. A molecular phylogenetic approach to Longidoridae (Nematoda: Dorylaimida). Nematology 2005; 7(1): 111-24.
[http://dx.doi.org/10.1163/1568541054192108]

[102] Mehdizadeh S, Shokoohi E. The genera *Nothacrobeles* Allen & Noffsinger, 1971 and *Zeldia* Thorne, 1937 (Nematoda: Rhabditida: Cephalobidae) from southern Iran, with description of *N. abolafiai* sp. n. Zootaxa 2013; 3637(3): 325-40.
[http://dx.doi.org/10.11646/zootaxa.3637.3.5] [PMID: 26046200]

[103] Blaxter ML, De Ley P, Garey JR, *et al.* A molecular evolutionary framework for the phylum Nematoda. Nature 1998; 392(6671): 71-5.
[http://dx.doi.org/10.1038/32160] [PMID: 9510248]

[104] De Luca F, Troccoli A, Duncan LW, *et al. Pratylenchus speijeri* n. sp. (Nematoda: Pratylenchidae), a new root-lesion nematode pest of plantain in West Africa. Nematology 2012; 14(8): 987-1004.
[http://dx.doi.org/10.1163/156854112X638424]

[105] Sheybani M. Biocontrol of Meloidogyne species associated with pistachio in Sirjan, south of Iran, 2014.

[106] Setterquist RA, Smith GK, Jones R, Fox GE. Diagnostic probes targeting the major sperm protein gene that may be useful in the molecular identification of nematodes. J Nematol 1996; 28(4) (Suppl.): 414-21.
[PMID: 11542511]

[107] Skantar AM, Carta LK. Molecular characterization and phylogenetic evaluation of the hsp90 gene from selected nematodes. J Nematol 2004; 36(4): 466-80.
[PMID: 19262827]

[108] Kovaleva ES, Subbotin SA, Masler EP, Chitwood DJ. Molecular characterization of the actin gene from cyst nematodes in comparison with those from other nematodes. Comp Parasitol 2005; 72(1): 39-49.
[http://dx.doi.org/10.1654/4138]

[109] Scholl EH, Bird DM. Resolving tylenchid evolutionary relationships through multiple gene analysis derived from EST data. Mol Phylogenet Evol 2005; 36(3): 536-45.
[http://dx.doi.org/10.1016/j.ympev.2005.03.016] [PMID: 15876542]

[110] Bik HM, Porazinska DL, Creer S, Caporaso JG, Knight R, Thomas WK. Sequencing our way towards understanding global eukaryotic biodiversity. Trends Ecol Evol 2012; 27(4): 233-43.
[http://dx.doi.org/10.1016/j.tree.2011.11.010] [PMID: 22244672]

[111] Lallias D, Hiddink JG, Fonseca VG, *et al.* Environmental metabarcoding reveals heterogeneous drivers of microbial eukaryote diversity in contrasting estuarine ecosystems. ISME J 2015; 9(5): 1208-21.

[http://dx.doi.org/10.1038/ismej.2014.213] [PMID: 25423027]

[112] Shokoohi E, Abolafia J. Redescription of a predatory and cannibalistic nematode, *Butlerius butleri* Goodey, 1929 (Rhabditida: Diplogastridae), from South Africa, including its first SEM study. Nematology 2021; 23(9): 969-86.
[http://dx.doi.org/10.1163/15685411-bja10089]

[113] Lobo I. Basic Local Alignment Search Tool (BLAST). Nature Education 2009; 1(1): 215.

[114] Nei M, Kumar S. Molecular evolution and phylogenetics 2000.
[http://dx.doi.org/10.1093/oso/9780195135848.001.0001]

[115] Tamura K, Stecher G, Kumar S. MEGA11: Molecular Evolutionary Genetics Analysis Version 11. Mol Biol Evol 2021; 38(7): 3022-7.
[http://dx.doi.org/10.1093/molbev/msab120] [PMID: 33892491]

[116] Shokoohi E. First observation on morphological and molecular characters of *Bitylenchus ventrosignatus* (Tobar Jiménez, 1969) Siddiqi, 1986 isolated from tomato in Dalmada, South Africa. Biologia (Bratisl) 2023; 78(12): 3599-607.
[http://dx.doi.org/10.1007/s11756-023-01494-4]

[117] Quist CW, Smant G, Helder J. Evolution of plant parasitism in the phylum Nematoda. Annu Rev Phytopathol 2015; 53(1): 289-310.
[http://dx.doi.org/10.1146/annurev-phyto-080614-120057] [PMID: 26047569]

[118] Qing X, Wang M, Karssen G, Bucki P, Bert W, Braun-Miyara S. PPNID: a reference database and molecular identification pipeline for plant-parasitic nematodes. Bioinformatics 2020; 36(4): 1052-6.
[http://dx.doi.org/10.1093/bioinformatics/btz707] [PMID: 31529041]

[119] Qing X, Wang Y, Lu X, *et al.* NemaRec: A deep learning-based web application for nematode image identification and ecological indices calculation. Eur J Soil Biol 2022; 110103408
[http://dx.doi.org/10.1016/j.ejsobi.2022.103408]

[120] Danovaro R, Gambi C, Della Croce N. Meiofauna hotspot in the Atacama Trench, eastern South Pacific Ocean. Deep Sea Res Part I Oceanogr Res Pap 2002; 49(5): 843-57.
[http://dx.doi.org/10.1016/S0967-0637(01)00084-X]

[121] Della Croce N, Albertelli G, Danovaro R, *et al.* Atacama Trench International Expedition (ATIE) Agor 60 Vidal Gormaz I Report 1998; 1-24.

[122] Baermann G. Eine einfache Methode zur Auffindung von Ankylostomum (Nematoden) larven in Erdproben. Geneeskundig Tijdschrift voor Nederlandsch Indië 1917; 57: 131-7.

[123] Seinhorst JW. On the killing, fixation and transferring to glycerin of nematodes. Nematologica 1962; 8(1): 29-32.
[http://dx.doi.org/10.1163/187529262X00981]

[124] Rowell DL. Soil Science: Methods and Applications 1994.

[125] De Waele D, Jordaan EM. Plant-parasitic nematodes on field crops in South Africa. 1. Maize. Rev Nématol 1998; 11(1): 65-74.

[126] Bolton C, De Waele D, Loots GC. Plant-parasitic nematodes on field crops in South Africa 3: Sunflower. Rev Nématol 1989; 12: 69-76.

[127] De Waele D, McDonald AH. Diseases caused by nematodes. 2000.

[128] Zeng Y, Ye W, Bruce Martin S, Martin M, Tredway L. Diversity and occurrence of plant-parasitic nematodes associated with golf course turfgrasses in north and South Carolina, USA. J Nematol 2012; 44(4): 337-47.
[PMID: 23482422]

[129] Colwell RK. Biodiversity: concepts, patterns and measurement 2009; 257-63.

[130] Sieriebriennikov B, Ferris H, de Goede RGM. NINJA: An automated calculation system for

nematode-based biological monitoring. Eur J Soil Biol 2014; 61: 90-3.
[http://dx.doi.org/10.1016/j.ejsobi.2014.02.004]

[131] Shokoohi E. Impact of agricultural land use on nematode diversity and soil quality in Dalmada, South Africa. Horticulturae 2023; 9(7): 749.
[http://dx.doi.org/10.3390/horticulturae9070749]

[132] Segata N, Waldron L, Ballarini A, Narasimhan V, Jousson O, Huttenhower C. Metagenomic microbial community profiling using unique clade-specific marker genes. Nat Methods 2012; 9(8): 811-4.
[http://dx.doi.org/10.1038/nmeth.2066] [PMID: 22688413]

[133] Gendron EM, Sevigny JL, Byiringiro I, Thomas WK, Powers TO, Porazinska DL. Nematode mitochondrial metagenomics: A new tool for biodiversity analysis. Mol Ecol Resour 2023; 23(5): 975-89.
[http://dx.doi.org/10.1111/1755-0998.13761] [PMID: 36727264]

[134] Rodrigue S, Materna AC, Timberlake SC, *et al.* Unlocking short read sequencing for metagenomics. PLoS One 2010; 5(7)e11840
[http://dx.doi.org/10.1371/journal.pone.0011840] [PMID: 20676378]

[135] Treonis AM, Unangst SK, Kepler RM, *et al.* Characterization of soil nematode communities in three cropping systems through morphological and DNA metabarcoding approaches. Sci Rep 2018; 8(1): 2004.
[http://dx.doi.org/10.1038/s41598-018-20366-5] [PMID: 29386563]

[136] Kenmotsu H, Ishikawa M, Nitta T, Hirose Y, Eki T. Distinct community structures of soil nematodes from three ecologically different sites revealed by high-throughput amplicon sequencing of four 18S ribosomal RNA gene regions. PLoS One 2021; 16(4)e0249571
[http://dx.doi.org/10.1371/journal.pone.0249571] [PMID: 33857177]

[137] Ren Y, Porazinska DL, Ma Q, Liu S, Li H, Qing X. A single degenerated primer significantly improves COX1 barcoding performance in soil nematode community profiling. Soil Ecol Lett 2024; 6(2)230204
[http://dx.doi.org/10.1007/s42832-023-0204-4]

[138] Shokoohi E, Mashela PW, Machado RAR. Bacterial communities associated with *Zeldia punctata*, a bacterivorous soil-borne nematode. Int Microbiol 2022; 25(1): 207 16.
[http://dx.doi.org/10.1007/s10123-021-00207-8] [PMID: 34553287]

[139] Shokoohi E, Machado RAR, Masoko P. Bacterial communities associated with *Acrobeles complexus* nematodes recovered from tomato crops in South Africa. PLoS One 2024; 19(6)e0304663
[http://dx.doi.org/10.1371/journal.pone.0304663] [PMID: 38843239]

[140] Dirksen P, Marsh SA, Braker I, *et al.* The native microbiome of the nematode *Caenorhabditis elegans*: gateway to a new host-microbiome model. BMC Biol 2016; 14(1): 38.
[http://dx.doi.org/10.1186/s12915-016-0258-1] [PMID: 27160191]

[141] Vafeiadou AM, Derycke S, Rigaux A, De Meester N, Guden RM, Moens T. Microbiome differentiation among coexisting nematode species in estuarine microhabitats: a metagenetic analysis. Nascimento FJA, editor. Front Mar Sci 2022; 9.
[http://dx.doi.org/10.3389/fmars.2022.881566]

[142] European Union Reference Laboratory for plant-parasitic nematodes. Protocols for culturing Plant-Parasitic Nematodes. Available from: https://sitesv2.anses.fr/en/system/files/Protocol_closed_containers_cystsMeloidogyne_EURL%20%282%29_1.pdf

[143] Kranse OP, Ko I, Healey R, *et al.* A low-cost and open-source solution to automate imaging and analysis of cyst nematode infection assays for *Arabidopsis thaliana.* Plant Methods 2022; 18(1): 134.
[http://dx.doi.org/10.1186/s13007-022-00963-2] [PMID: 36503537]

[144] Coyne DL, Adewuyi O, Mbiru E. Protocol for in vitro culturing of lesion nematodes: Radopholus similis and Pratylenchus spp on carrot discs 2014.

[145] Imamah AN, Supramana S, Damayanti TA. In vitro cultivation of *Aphelenchoides besseyi* Christie on fungal cultures. Jurnal Perlindungan Tanaman Indonesia 2020; 24(1): 43-7.
[http://dx.doi.org/10.22146/jpti.42227]

[146] Zhang Y, Wen TY, Wu XQ, Hu LJ, Qiu YJ, Rui L. The *Bursaphelenchus xylophilus* effector BxML1 targets the cyclophilin protein (CyP) to promote parasitism and virulence in pine. BMC Plant Biol 2022; 22(1): 216.
[http://dx.doi.org/10.1186/s12870-022-03567-z] [PMID: 35473472]

[147] Koohkan M, Shokoohi E. Mass Production and Prey Species of *Mylonchulus sigmaturus* (Nematoda: Mylonchulidae) in the Laboratory. Acta Zool Bulg 2014; 66(4): 555-8.

[148] Ayub F, Seychelles L, Strauch O, Wittke M, Ehlers RU. Monoxenic liquid culture with Escherichia coli of the free-living nematode Panagrolaimus sp. (strain NFS 24-5), a potential live food candidate for marine fish and shrimp larvae. Appl Microbiol Biotechnol 2013; 97(18): 8049-55.
[http://dx.doi.org/10.1007/s00253-013-5061-0] [PMID: 23812335]

[149] Stiernagle T. Maintenance of *C. elegans*, Worm Book, ed. The *C. elegans* Research Community, Woorm Book 2006.

[150] Brenner S. The genetics of *Caenorhabditis elegans*. Genetics 1974; 77(1): 71-94.
[http://dx.doi.org/10.1093/genetics/77.1.71] [PMID: 4366476]

[151] Yemini E, Kerr RA, Schafer WR. Preparation of samples for single-worm tracking. Cold Spring Harb Protoc 2011; 2011(12)pdb.prot066993
[http://dx.doi.org/10.1101/pdb.prot066993] [PMID: 22135667]

[152] Lewis JA, Fleming JT. Basic culture methods. Methods Cell Biol 1995; 48: 3-29.
[http://dx.doi.org/10.1016/S0091-679X(08)61381-3] [PMID: 8531730]

[153] Joshi I, Kumar A, Singh AK, *et al.* Development of nematode resistance in Arabidopsis by HD-RNA--mediated silencing of the effector gene Mi-msp2. Sci Rep 2019; 9(1): 17404.
[http://dx.doi.org/10.1038/s41598-019-53485-8] [PMID: 31757987]

[154] Wang G, Weng L, Li M, Xiao H. Response of gene expression and alternative splicing to distinct growth environments in tomato. Int J Mol Sci 2017; 18(3): 475.
[http://dx.doi.org/10.3390/ijms18030475] [PMID: 28257093]

[155] Munns David PD. Controlling the Environment: The Australian phytotron, the Colombo plan, and postcolonial science. British scholar 2010; 2 (2): 197–226.
[http://dx.doi.org/10.3366/brs.2010.0203]

[156] Biotech Ch. Phytotron and Greenhouse 2021. Available from: https://www.chbio.com.tw/en/global_rd_center/phytotron_greenhouse

[157] Quesenberry KH, Baltensperger DD, Dunn RA, Wilcox CJ, Hardy SR. Selection for tolerance to root-knot nematodes in red clover. Crop Sci 1989; 29(1): 62-5.
[http://dx.doi.org/10.2135/cropsci1989.0011183X002900010014x]

[158] Snyder DW, Opperman CH, Bird DM. A method for generating *Meloidogyne incognita* males. J Nematol 2006; 38(2): 192-4.
[PMID: 19259447]

[159] Windham GL, Williams WP. Host suitability of commercial corn hybrids to *Meloidogyne arenaria* and *M. incognita*. J Nematol 1987; 19(Annals 1): 13-6.
[PMID: 19290266]

SUBJECT INDEX

A

Acid 79, 81
  deoxyribonucleic 79
  nucleic 81
Activity 70, 75
  exonuclease 75
  highest enzyme 70
AFLP 71, 72, 73
  analysis 73
  -PCR 71, 72
  technique 71, 73
Agarose gel test 77
Agricultural 1, 2
  production 2
  productivity 1
Agroecosystems 3
Alcohol burner 33
Alignment function 97
Amplification process 80
Amplified 58, 71, 73
  fragment length polymorphism (AFLP) 58,
    71, 73
  length fragment polymorphism 71
Ancestors, hypothetical common 89
Animals, microscopic 117
Aphelenchoides 27, 70, 111, 112
  individuals 112
  surface 111
Approaches, rDNA metagenomic 105
Ascariasis 1
Ascaris infections 1
Automated systems 73

B

Bacterial infections 1
Bacterivore footprint 103
Baermann Funnel technique 13, 99
Basal bulb 47, 48, 49, 50, 52
  length 47, 48, 49, 50, 52
  width 47, 48, 49, 50, 52

Basal Index 102, 103
Bases, matching 89
Basic local alignment search tool 86
Baunacke method 19, 26
Bayesian tree 92
Beaker, plastic 20
Biodiversity 11, 13, 91, 98, 100, 108
  indices 100
Bioinformatics 81, 86, 97
  methods 97
Bitylenchus.mas 91
Blast 86, 88, 90
  algorithm 86
  tool 90
Blender method 20
Buccal cavity 51
  length 51
  width 51

C

Carbon dioxide 55
Chikungunya 23
Cinerea mycelia 112
Cloning, blunt-ended PCR 75
Community indices 98
Computational process 94
Concentration 23, 55, 77, 82, 84
  measures protein 77
Condensation water droplets 25
Constructing trees 92
Containing paraffin wax 34
Crop(s) 1, 2, 3, 4, 11
  agricultural 2
  pests 1
Culturing nematodes 117
Cyst nematodes 2, 18, 67, 109, 110, 113

D

Damage, mechanical 5
Debris material 15

DIC microscope 41, 42
Differential interference contrast (DIC) 36, 41, 43, 57
   microscopy 36
Distance 91, 92, 93
   genetic 91
   methods 92, 93
DNA 58, 60, 61, 63, 64, 70, 71, 74, 76, 77, 78, 80, 81, 82, 84, 89, 98
   complementary 74
   double-strand 74
   fragments range 71
   impure 84
   markers 58
   mitochondrial 98
   plasmid 74
   polymerase 81
   -protein complexes 60
   ribosomal 98
   sequencing method 81
   target 74
   visualization techniques 58
DNA extraction 17, 58, 59, 60, 61, 64, 68, 80
   kit 60, 68, 80
   methods 80
   process 60
DNA fingerprinting 68, 71
   method 71
DNA template 61, 68, 76, 77, 82, 83, 84
   contaminated 83
   target nematode's 76
DNAzol isolation reagent 63

E

Efron's bootstrap resampling method 96
Electrical conductivity (EC) 99
Electron 29, 56
   microscopy 56
   transmission 29
Electronic 45
   column 45
   console 45
Electrophoresis 78
Elephantiasis 1
Entomopathogenic nematode (EPN) 21, 23, 27, 28, 109, 117
Environment 1, 3, 108, 112, 113, 117
   artificial 112
   controlled greenhouse 117

Environmental DNA analysis 81
Enzyme(s) 70, 71
   activities 70
   restriction 70, 71

F

Felsenstein's bootstrap test 96
Fever, yellow 23
Food 1, 2
   products 2
   safety measures 1

G

Gel electrophoresis 68, 84
Gene(s) 81, 94, 113
   expression 81, 113
   taxonomic marker 94
Genomic DNA 59, 74, 75, 76, 78
   extraction 59
Global 2, 11, 21, 99
   economic losses 2
   positioning system (GPS) 11, 21, 99
Glycerine, anhydrous 99
Greenhous(s) 109, 110, 113, 114
   plastic 113
   technique 114
Ground wheat rusk 24
Growth 4, 110, 117
   plant 4, 110
Gubernaculum length 49, 50
Gymnostom width 50

H

Health, environmental 1
Hydroponic 116
   system 116
   tank 116

I

Images, analyze nematode 97
Imaging techniques 57
Immersion oil 41
Incubation method 16
Index 100, 101, 103
   measures 101

Shannon 100, 101, 103
Indian populations of pigeon pea cyst
    nematode 73
Indices, functional 102
Infection 1, 2, 23
    parasitic 1
    virus 2
Isolation of genomic DNA 59

L

Lens 37, 40
    ocular 40
    oil immersion 37

M

Media, solid culture 109
Median bulb width 46
Meloidogyne 17, 20, 27, 68, 70, 79, 102, 109,
    110, 114
    and cyst nematodes 109
Metagenomic analysis 98, 105
Microbial balance 3
Microscope 29, 36, 37, 38, 39, 40, 41, 45, 55,
    56, 57
    scanning electron 55, 56
    transporting 41
Microscopy, scanning 56
Mistifier technique 27, 28
Molecular 58, 70, 80, 110
    biology method 70
    taxonomy 80, 110
    techniques 58
Monoxenic cultures 112

N

Nail polish 32, 33
    colorless 32
    transparent 33
NCBI websites 90
Nematoda 95, 104, 117
    behavior 117
Nematodes 1, 2, 4, 9, 11, 16, 17, 20, 21, 23,
    24, 25, 28, 31, 36, 41, 43, 55, 57, 58, 59,
    62, 80, 97, 99, 103, 104, 105, 108, 109,
    111, 112, 117
    bacterivorous 11
    community indices 103

cylinder 111
dehydration 55
digestion 62
endoparasitic 20
freshwater 9
fying plant-parasitic 97
genes 58
growth medium (NGM) 112
infestation 2
isolation 21, 58
live 28, 31, 109
mermithid 23, 24
metabarcoding analysis 105
metagenomics analysis 104
microbiome 105
sedentary 20
sedentary plant-parasitic 17
soil-dealing 21
sterilizing 111
suspension 23, 111
taxonomists 57
taxonomy 41
Nematode biodiversity 98, 99, 100, 102
    analysis 99, 100
    assessment 98
    indices 100
Nematode DNA 60, 63, 65, 76
    extraction 60, 63
Nematode populations 5, 9, 58, 116
    contaminating 5
Nematode species 9, 11, 12, 22, 23, 35, 41,
    70, 81, 107, 109
    cryophilic plant-parasitic 11
Nematologists 31, 36, 73, 75, 97, 108, 109,
    117
Nematology 11, 76, 99, 113
    laboratories 11, 76, 99
    plant 113
Neural networks 97
NGS sequencing 82
Nitrogen fixation 3
Nutritional disorders 2

O

Oblique contrast 41
Oostenbrink's reproduction factor 116

P

Parasitic nematode 1, 23
Parsimony principle 94
PCR 58, 61, 62, 68, 74, 75, 75, 76, 77
    amplification 74
    -based method 68
    -grade water 62
    machine block 61
    processing 58, 74, 75, 76, 77
    product on gel electrophoresis 68
    reactions 61, 76
    techniques 74
Pharyngeal gland 44
    dorsal 44
Photograph, ultraviolet illumination 69, 72
Phylogenetic(s) delves 89, 90, 92, 93, 94, 95, 96
    trees 89, 90, 92, 93, 94, 95, 96
Phytotron photograph 114
Pictorial measurement guide 45
Plant(s) 2, 4, 5, 9, 13, 15, 16, 17, 25, 28, 44, 45, 58, 98, 99, 113
    -feeding 28
    parts 17
    pathology 113
    infected 5
    -parasitic nematodes 2, 4, 5, 9, 13, 15, 44, 45, 58, 98, 99
    material 5, 16, 25
Plastic 6, 8, 13, 21, 25
    bags 6, 8
    containers 6, 8, 13, 21, 25
Plate, hot 33
Polymerase 58, 65, 68, 71, 73, 75, 76, 77, 80, 84
    chain reaction 73, 77, 80
    change reaction (PCR) 58, 65, 68, 71, 73, 75, 76, 80, 84
Polymorphism 58, 68, 71
    amplified fragment length 58, 71
    bands 71
    restriction fragment length 68
Pooled nematodes 62
Population(s) 1, 2, 4, 12, 100
    densities fluctuate 4
    density 2, 100
    dynamics 12
    wild fish 1
Posterior ovary 44

Potato dextrose agar (PDA) 111
Primer(s) 75, 84
    -extension reactions 75
    oligonucleotide 84
Processing, mitochondrial DNA 76
Prominence value (PV) 100
Protein 60, 76, 77, 86, 90, 103
    sperm 76
Proteinase 60, 61, 62, 63, 64
    deactivate 61
Pseudacrobeles macrocystis 86
Pyrosequencing method 82

R

Research 1, 2, 4, 13, 28, 36, 58, 76, 107, 109, 117
    agricultural 2, 4
    biological 117
    molecular biology 28
    nematological 12, 36, 76
Restriction fragment length polymorphism (RFLP) 68
RFLP technique 69
Rhabditida 11, 45
    free-living 45
Romanomermis 23, 24
    nematodes 24
Root(s) 111, 114
    infected 111, 114
    -knot inoculation 114
Rotten woods 20

S

Samples 4, 5, 6, 9, 11, 12, 18, 21, 55, 58, 63, 80, 91, 98, 99, 101, 102, 103
    collected 12
    combined 9
    complex 103
    composite 99
    mixed 5
    random 21
    sediment 98
    small 80
Sand, clean 110
Scanning electron microscopy 36, 45, 57
SCAR 65, 66, 67
    marker 65, 67
    primers 67

ultraviolet illumination photograph 66
Seasonal fluctuation 99
Seeds, soak tomato 116
Segments, expansion 76
Seinhorst 18, 27
   cyst extraction elutriator method 18
   elutriator density 27
Sequence(s) 11, 58, 65, 80, 81, 82, 86, 87, 89,
    90, 94, 95, 96
   characterized amplified region (SCAR) 58,
    65
   cropping 11
   genetic 80
   next-generation 81
Software 89
   computer 89
   tools 89
Soil 1, 2, 3, 4, 5, 6, 9, 11, 13, 14, 15, 19, 21,
    28, 98, 99, 105, 107, 108, 111, 114
   and plant samples 11
   biodiversity 107
   communities 98
   dried 19
   dry 5
   ecosystems 98, 107
   fertility problems 2
   health 3, 108
   mixing 5
   moist 21
   nematodes 28
   nutrition 3
   profile 9
   rhizosphere 9
   sediments 98
   sterilized 114
   systems 2
Soybean cyst nematode (SCN) 65
Sterilized coconut coir fibers 25

T

Taq 73, 75
   DNA polymerase 75
   polymerase 73
Taxonomical 9, 33, 36, 43
   examination 36
   purposes 9, 43
Thermos shaker 64
Thermostable DNA polymerases 75
Tools 5, 6, 57, 80, 86, 93, 97, 105, 117

comprehensive analytical 93
   indispensable 57, 105
Transfer 59, 116
   nematodes 59
   plants 116
Tray method 13, 14, 15, 20, 25, 27, 28
Tree 2, 9, 11, 92, 94, 96
   -building method 96
   condensed 92
   evolutionary 94

W

Water 2, 14, 15, 18, 19, 23, 24, 25, 29, 30, 31,
    32, 55, 59, 65, 99, 110, 111
   deionized 59
   double-distilled 110
   hot 15, 65
   removing redundant 30
   sterilized 24, 111
   stress 2
   thin layer of 14, 29, 30
Wood and compost extraction 20

www.ingramcontent.com/pod-product-compliance
Lightning Source LLC
Chambersburg PA
CBHW041714210326
41598CB00007B/652